SpringerBriefs in Applied Sciences and Technology

Thermal Engineering and Applied Science

Series Editor

Francis A. Kulacki, University of Minnesota, Minnesota, USA

W0115098

For further volumes:
http://www.springer.com/series/10305

Navdeep Singh · Debjyoti Banerjee

Nanofins

Science and Applications

 Springer

Navdeep Singh
Debjyoti Banerjee
Department of Mechanical Engineering
Texas A&M University
College Station, TX
USA

ISSN 2193-2530   ISSN 2193-2549 (electronic)
ISBN 978-1-4614-8531-5   ISBN 978-1-4614-8532-2 (eBook)
DOI 10.1007/978-1-4614-8532-2
Springer New York Heidelberg Dordrecht London

Library of Congress Control Number: 2013950084

Printed on acid-free paper

Springer is part of Springer Science+Business Media (www.springer.com)

Preface

The topics in this book relate to some of the materials covered in the graduate course Multi-Phase Flows and Heat Transfer (MEEN 624), a PhD-level elective course which I have been teaching at Texas A&M University for the past 8 years. This book is meant to formalize the several interesting topics and stimulating discussions I (D. B.) had with my students for this class using the "Peer Teaching Concepts."

The discipline of multi-phase flows and heat transfer, as well as, thermofluids in general, have traditionally focused on large-scale devices and systems while developing the fundamental concepts based on molecular scale interactions for generating the equations for conduction, convection, and radiation. Multidisciplinary studies of these traditional subjects involving nanotechnology have led to surprising discoveries and counter-intuitive results. A number of good books have been published in the past two decades involving micro/nanoscale heat transfer (e.g., Majumdar, Peterson, etc.) as well as micro/nanofluidics. The scope of the topics addressed in these books has been limited to mostly conduction in thin films and radiation heat transfer in small devices or bio-chemical reactions in flow conduits (e.g., millimeter-scale, microscale, and nanoscale).

A textbook that addresses the fundamental advances in transport phenomena due to application of nanotechnology is currently lacking in the two-phase flow literature. To complement my lecture material, I (D. B.) involved the students in my class by assigning a term-project where they would cull the literature for papers on a particular topic and with my guidance teach their peers on these advanced topics of contemporary relevance as well as identify and justify future directions; these topics are expected to evolve into—based on the current limitations. This approach is based on the "Peer Teaching Concepts" where the students are known to learn better by teaching their peers while the instructor serves as a moderator and guide. Often, I had to refer the students working on a particular topic to several texts and papers, each of which assumes that the reader is aware of the nuances of the specialized contents of their erudition. This realization of the need for a textbook that comprehensively collects, summarizes, and discusses the implications of nanotechnology in multi-phase flows provided the impetus to develop this text. The text is meant to support the graduate engineering community and researchers with a coherent textbook to enable them to rapidly immerse into the nuances of the perturbation of the transport phenomena and their effects on a much larger scale. This book is meant for a broad audience in engineering and science. It is also suitable for self-study—provided the reader has sufficient knowledge of fluid mechanics and heat transfer.

Recent advances in nanotechnology have enabled the engineering of surface structures with well-defined nanometer scale tolerances ("nanostructures"). This was enabled by the development of metrology techniques for characterizing the nanometer scale (or "nanoscale") features with sub-micron precision for both positioning and measurements. Synthesis techniques for realizing novel nanomaterials and the sophisticated materials characterizing techniques have also enabled the application of nanotechnology to thermofluids research.

At the heart of all these reports in the literature is the formation of "nanofins" and the associated peculiarities in transport phenomena, termed here as the "nanofin effect." After culling through the research results in the literature, we have come to the conclusion that the nanofin effect is ubiquitous in all the studies involving the application of nanotechnology in thermofluids research. Experimental studies have shown that the nanofin effect is responsible for the enhancement in the heat flux in pool boiling, convection of nanofluids, and the specific heat capacity (as well as thermal conductivity) of nanofluids. These experiments also show that heaters containing organic nanofins (e.g., carbon nanotubes or "CNT") provide lower levels of heat transfer enhancement than that for silicon nanofins. This can be attributed to the modulation of the total thermal resistance (or thermal impedance for unsteady situations) to the heat flow from nanostructures to the surrounding fluid molecules. Value of the total thermal impedance in the network is estimated to be dominated by the value of the interfacial thermal resistance ("Kapitza Resistance," R_k). So the variation of the interfacial thermal resistance needs to be explored for various scenarios in order to determine the overall effectiveness of a nanofin. Similarly, adsorption of the fluid molecules (solvent phase) on the surface of the nanofins is expected to create an extra phase in the system owing to their higher density of packing (leading to different values of material properties of the mixture—such as thermal conductivity, specific heat capacity, rheology, chemical potential, etc.).

It is expected that the treatment of these recent developments in thermofluids research involving application of nanotechnology from the perspective of the nanofin effect will help to demystify this subject area for graduate students, entrants into this research topic as well as for practitioners in this field.

Organization of the Book

Chapter 1 provides an overview of the subject material covered in this book—which includes definitions, engineering motivation/justification for this research topic, historical perspective, and links to the engineering/technology application landscape. Chapter 2 provides a background on the carbon nanotubes, its novel properties and present as well as future applications. Next, Molecular Dynamics simulations are discussed. This chapter also describes the velocity verlet method used to numerically integrate the equations of motion and the polymer consistent force field (PCFF) used in the present study. At the end of the chapter, the literature on interfacial thermal resistance is reviewed. Simulation results are presented and discussed in Chap. 3. Chapter 4 summarizes the results and identifies the future scope of work.

Caveat

In typing the several equations in this book, it is possible that typographical errors may have occurred. We tried to detect and eliminate typing, spelling, grammatical, and other errors, but we may have overlooked some that remain to be detected by the readers. We will sincerely appreciate any feedback from the reader in notifying us of the potential mistakes (typographical or otherwise); and our electronic contact information is provided in this book. We also welcome any comments or suggestions regarding the improvement of the future editions of this book and also any suggestions for other book titles to complement this endeavor.

Acknowledgments

We (N. S. and D. B.) would like to thank several individuals and institutions.

First and foremost, I (D. B.) would like to thank the honorable Vice-Chancellor (Dr. Souvik Bhattacharyya, formerly Dean at Indian Institute of Technology/I.I.T. at Kharagpur) of Jadavpur University (J.U.) and faculty members of the nanotechnology program at J.U., especially Profs. Manoj Mitra and Kalyan Chattopadhyay (as well as their graduate students). I was working on the book manuscript at the time of my sabbatical while organizing the "1st International Workshop on Nanomaterials (IWoN): Engineering Photon and Phonon Transport" at J.U. with financial support from the V-C of J.U., the U.S. Air Force Office of Scientific Research (AFOSR-AOARD) and the U.S. Office of Naval Research—Global (ONR-G). The cooperation and help from the faculty at J.U. as well as the workshop sponsors during my sabbatical enabled me to find the time to work on the book manuscript. In this regard—I (D. B.) would like to thank Drs. Rengasammy Ponnappan (AFOSR) and Ingrid Wysong (Asian Office of Aerospace Research and Development/ AOARD, Tokyo, Japan) as well as Dr. Gabriel Roy (ONR-G, Singapore) for the workshop support. I (D. B.) also thank the Department of Mechanical Engineering (MEEN) at Texas A&M University (TAMU) for supporting my sabbatical—nee Faculty Development Leave (FDL) during the Fall Semester of 2012 that enabled me to focus on finishing the writing of this book. We also acknowledge Prof. Frank A. Kulacki at the University of Minnesota for inviting us to write this book.

We also acknowledge the support from various agencies, listed below, that provided us the opportunity to create the scientific/technical knowledge that was gained from performing various research projects—which led to the comprehension/development of the philosophy underlying the term "nanofin effect" and eventually the writing of this book:

1. National Science Foundation (NSF): CBET-TTTP and SBIR programs
2. Department of Energy (DOE): Solar Energy Technology Program
3. U.S. Air Force: AFRL, AFOSR, AOARD, and SBIR programs
4. ONR: ONR-G, SPAWAR, and SBIR programs
5. NASA: TiiMS and URETI programs
6. Texas Space Grants Consortium (TSGC): New Investigator Program (NIP)
7. Texas A&M Engineering Experiment Station (TEES)
8. Qatar Foundation: QNRF program
9. DARPA: MTO, Micro/Nano Fluidics Fundamentals Focus (MF^3) center at the University of California at Irvine and corporate level sponsors of MF^3 including:

   a. Beckman Coulter Inc.
   b. BIOCOM
   c. CFDRC
   d. Douglas Scientific
   e. ESI Group
   f. Invitrogen
   g. IDEX Health and Science
   h. Lawrence Livermore National Labs (LLNL)
   i. Life Technologies (formerly Applied Biosystems/Applera)
   j. Microfluidics Innovations
   k. Monsanto Company
   l. NASA Ames Research Center
   m. Pioneer Hi-Bred International Inc.
   n. SHRINK Nanotechnologies
   o. Sierra Proto Express
   p. Symbient Product Development

10. Corporate level collaborators/sponsors for our research group—Multi-Phase Flows and Heat Transfer Laboratory (MFHTL):

    a. 3M Corp Innovation Center
    b. NanoInk Inc.
    c. NanoMEMS Research LLC
    d. Aspen Thermal Systems
    e. Lynntech Inc.
    f. ADA Technologies
    g. General Dynamics (Anteon Corporation)
    h. General Electric (G. E.) Corporate Research
    i. Irvine Sensors Corp.

j. Trianja Inc. (Photronics Corp.)
k. Silicon Values Partners (B G Group)

The sponsor support led to the publication of M.S. and Ph.D. thesis—contents of which have contributed directly or indirectly to the materials discussed in this book. Hence, we would like to acknowledge the past M.S. and Ph.D. students who graduated from our research group. The students include:

I. Ph.D. Students: Hee-Seok Ahn (2007), Vijaykumar Sathyamurthi (2009), Dong-hyun Shin (2011), Sae-il Jeon (2011), Seok-won Kang (2012), Seung-hwan Jung (2012), Byeongnam Jo (2012), and Jiwon Yu (2012).
II. M.S. Students: J. A. Rivas-Cordona (2005), N. Sinha (2005), Rupakula V. (2005), V. Sathyamurthi (2006), I. C. Nelson (2007), D. Huitink (2007), S. Shenoy (2007), S. Sriraman (2008), R. Banneyake (2008), S. Datta (2008), R. Gargate (2008), S. Gauntt (2009), S. Glenn (2009), Y. Lin (2010), S. Hansen (2012), and N. Niedbalski (2012).

In the interest of brevity—we apologize in advance for omitting the names of the collaborators and various experts whose advice and comments have enriched our understanding of the transport phenomena at the nanoscale and have also directly or indirectly influenced the writing of this book.

Last, but not the least, our special thanks go to our family members for their support for this endeavor. N. S. would like to thank Harjinder in this regard. D. B. would like to thank his spouse Dolan as well as children (Debkonya and Devkumar) in this regard.

College Station, TX, USA, May 30, 2013   Navdeep Singh
                                                                Debjyoti Banerjee

Contents

Chapter 1
Introduction

Abstract Some of the ubiquitous terminologies used in this book are defined initially, as a caveat and to allay potential confusion. A considerable portion of this book is based on contents from a previous publication (Singh N, 2010, PhD Thesis, Texas A&M University). The nanofin effect and the associated nuances of this phenomenon are introduced in this chapter.

Terminologies, Definitions, and Conundrums

"Nanotechnology" is the discipline that pertains to utilization of matter and shapes with *geometric dimensions* in the range of ~ 1–10^3 nm (where 1 nm = 10^{-9} m). "Nanostructure" refers to shapes—which have at least one geometric feature (e.g., length, diameter, pitch, gaps between structures) that is "less than ~ 100 nm" (in some instances, this criterion has been relaxed in the literature and loosely referred to be "less than ~ 500 nm").[1] The term "nanostructure" is a generic term for "nanoparticles" (NP), "nanotubes" (NT), and "nanowires" (NW). Sometimes the term NW is loosely used in the literature to refer to cylindrical-shaped NP that is more than 100 nm in diameter (while NT usually means that the cylindrical-shaped NP has a diameter less than 100 nm).

When NP are *deliberately* coated on a surface (by physical or chemical deposition or using in situ synthesis techniques), it is called "nanocoatings." Sometimes unintended precipitation of NP from nanofluids on the conduit walls can also effectively cause nanocoatings to be formed, especially on heat exchanging surfaces—however, that is not usually a deliberate action for realizing nanocoatings. In contrast, "nanostructures" can be obtained by sculpting the

---

[1] Similarly, in boiling literature, "microlayer" refers to molecular structures of the nonvapor phase (liquid or semisolid structures) that are estimated to be ~ 100 nm thick. "Microparticles" and "microstructures" refer to features with dimensions in the range of 1–500 microns (μm), where 1 μm = 10^{-6} m.

N. Singh and D. Banerjee, *Nanofins*,
SpringerBriefs in Thermal Engineering and Applied Science,
DOI: 10.1007/978-1-4614-8532-2_1, © The Author(s) 2014

surface (e.g., by etching), often to obtain precisely engineered features—an act that is often termed as "nanofabrication" or "nanomanufacturing." In contrast, the act of obtaining nanocoatings, NP, and nanostructures without the requirement for precisely engineered features is termed as "nanosynthesis." Liquid solvents doped with a small concentration of NP (typically around 1 % or less) resulting in the formation of a stable colloidal suspension—are called "nanofluids." The solid phase of the nanofluids or mixtures of NP with any material in the solid state are termed as "nanocomposites."

Thermofluidics (or thermofluids) is the discipline that utilizes the principles of thermodynamics, fluid flow, heat and mass transfer (including radiation heat transfer as well as chemical and nuclear reactions) for the comprehensive study of the transport of mass and energy. At subnanometer length scales, laws governing equivalence of matter and energy apply. At length scales larger than ~ 1 nm, mass and energy can be treated as separate parameters that are governed by different *fundamental dimensions* (i.e., [M] and [ML^2 T^{-2}], respectively, where the fundamental dimensions are [Mass], [Length], [Time], etc.). The physical models that are commonly used for describing the transport of matter and energy are often coupled. The transport models yield nonlinear equations with time varying terms. The temporal variations of the solutions from these equations can be sensitive to small fluctuations in the values of the initial conditions (i.e., the solutions to these equations as applied to describing natural phenomena—can border on statistical or *nondeterministic* behavior—thus displaying "chaotic behavior" and "fractal structures"). Very often, physical models using force interactions (also called "mechanistic models") are utilized instead as an approximation—in order to derive a simplistic (i.e., *deterministic*) physical representation of these complex phenomena that involve the interactions between matter and energy in nature.

The term "fractal" refers to self-replicating patterns that retain their semblance of self-similarity at different length scales. The term "chaotic behavior" (or the principles of "chaos theory") refers to a repeating sequence of events in natural phenomena that do not return to exactly the same set of "initial configurations" but with each repetition of the cycle of events return to the vicinity of the set of "initial configurations." The key benefit of representing the behavior of a system using the principles of chaos theory is that the net (ensemble) behavior of a multiple parameter system can often be represented (i.e., the abstraction is represented) by the phase space plot of a single variable only, provided the phase plots are performed for a "sufficient" length of time (defining "sufficiency" for the size of the required data is a conundrum).

Despite the obvious short comings—a rich body of insightful approaches have been explored in the study of natural phenomena by utilizing mathematical tools from fractals/chaos theory. However, such efforts are still in their rudimentary stages in the thermofluidics literature as yet. These mathematical tools have barely been utilized to their fullest potential by the practitioners in the art and science of thermofluidics. This is primarily because the mathematical tools are yet to be developed to a mature level by mathematicians—so that behavioral analyses of discrete data sets (that are obtained from the studies in thermofluidics literature) can be performed in tractable manner using principles of chaos theory. Therefore, practical applications utilizing principles of chaos theory are yet to be realized in commercial thermofluidic devices (e.g., heat exchangers or thermohydraulic control of nuclear power plants). To wrap up this discussion on terminologies and prevent getting distracted from the primary focus (the topic) of this book—it is worth mentioning that the discourse of fractals and chaos theory in thermofluidics literature has a rich potential for further development.

Segments, Components, and Subliminal Liaisons

In the realm of nanotechnology, the contemporary research and development (R&D) efforts in thermofluidics can primarily be categorized into two broad topics (or approaches):

1. nanocoatings/nanostructures on heat exchanging surfaces and
2. nanoparticle additives in bulk material matrices (i.e., "nanofluids" and "nanocomposites").

The goal of this book is to show that the heat transfer enhancement (or sometimes degradation) reported in the literature for these two disparate approaches involving nanotechnology in thermofluidics—are subliminally connected since they are both dominated by the same phenomenon, termed here as the "nanofin effect."

"Nanofin effect" is a phenomenological term that can be used to rationalize the various behaviors (often contradictory) for these two disparate topics. The term "nanofin effect" encompasses a combination of several mechanisms, which includes surface adsorption of solvent molecules ("compressed layer") on solid surfaces, interfacial thermal resistance between the fin surface and the surrounding fluid molecules (also called "Kapitza resistance" which arises from the impedance mismatch of thermal/molecular level vibrations between two different materials in contact), chemical potential gradient-mediated thermal transport due to higher concentration of solvent species at the interface with the NP (i.e., on the surface of the NP), etc. The term "nanofin" will be defined in the next section.

Nanocoatings, Nanofluids, and Nanofins (Nano-this and Nano-that!)

In heat transfer pedagogy—the different theoretical constructs (e.g., equations and correlations for conduction, convection, and sometimes radiation) come together in the end when they are applied for design optimization of devices used in various commercial endeavors. The design optimization topics in heat transfer pedagogy are primarily twofold:

1. analysis of heat transfer from extended surfaces (called "fins") and
2. analysis of heat transfer in heat exchangers.

In conventional heat transfer pedagogy, the heat transfer in the presence of a fin (i.e., the thermal analogue of electrical current) for a finite temperature drop (i.e., the thermal analogue of electrical potential difference) is dependent on two thermal resistances:

(a) conduction resistance in the fin and
(b) convection resistance in the fluid.

Analysis of fins is primarily achieved through the fin equation:

$$\frac{d^2T}{dx^2} - \frac{hP}{kA_c}(T - T_\infty) = 0 \tag{1}$$

$$\frac{d^2\theta}{dx^2} - m^2\theta = 0 \tag{2}$$

where h is the heat transfer coefficient, P is the perimeter of the fin cross section (fin geometry), k is the thermal conductivity of the fin material, A_c is the cross-sectional area of the of fin geometry, θ is the temperature variable ($\theta = T - T_\infty$), where T is the temperature at any location in the fin, T_∞ is the bulk temperature of the fluid in contact with the fin, and m is called the fin parameter.

Implicit in the derivation of the fin equations is the assumption of continuity—i.e., the size of the fin is much larger than the molecular mean-free path of the fluid molecules in contact with the fin. However, if the geometric dimension of the fin is of the same order of magnitude as the molecular mean-free path (say, less than 500 nm for conventional applications), it can be termed as a "nanofin." For nanofin, the continuity assumptions are not valid. Hence, an extra thermal resistance needs to be accounted for—which is the interfacial resistance between the solid material of the fin and the fluid molecules in contact with the fin. The interfacial resistance is termed as the "Kapitza resistance" (R_k). Kapitza resistance is a material property that is determined by the choice of the material for both the fin and the fluid in contact. Hence, the extended surface analyses needs to be modified to account for this extra interfacial thermal resistance (and any associated physical phenomena such as chemical and physical adsorption interactions on the surface of the nanofin) and is termed as the "nanofin effect."

For steady-state operation of nanofins in single-phase flow—the heat transfer coefficient in the fin equation can be replaced with an effective heat transfer coefficient that captures the effect of Kapitza resistance as follows:

$$\frac{1}{h_{\text{eff}}} = \frac{1}{h} + \frac{1}{R_k} \tag{3}$$

However, for nonsteady operations or quasi-steady assumptions (which is usually the situation at the molecular level), additional terms need to be included in the transient analyses, such as the thermal (or the thermochemical) analogue of capacitative and inductive terms arising from the surface adsorption of the fluid molecules on the nanofin surface (e.g., due to van der Waals force and ionic force interactions). The governing equations in such a situation are contextual, and the transient analysis of the thermal impedance circuit is required for the particular configuration of nanofins (i.e., depending on the geometric layout/arrangement of the nanofin arrays). Next, it will be shown that the heat transfer from nanocoatings (as well as from surfaces in contact with nanofluids)—is primarily mediated through the formation of nanofins—called the "nanofin effect."

Since early twenty-first century, there have been several studies showing spectacular improvements in boiling heat transfer where the heater surface was coated with nanomaterials (such as carbon nanotubes, CNT, and metal nanowires, MNW). Similarly, bulk materials engineered to form surface nanostructures (e.g., silicon nanofins) showed even higher levels of heat flux enhancement during pool boiling. Initially, it was not clear from the experiments if the material properties of the nanomaterial coating (or the engineered surface nanostructures) are the dominant factor for heat transfer enhancement or if the shape (or surface area enhancement) of the resultant nanostructures in the nanocoatings plays a more dominant role (and therefore, the material properties of the NP play a secondary role). Nanofin analyses showed that silicon nanowires have $\sim 1{,}000$-fold lower values of Kapitza resistance than for organic materials (such as CNT). Thus, even if the thermal resistance within the silicon nanofin is higher than for CNT—the Kapitza resistance is substantially lower for silicon nanofins—thus resulting in a substantially lower value of the total thermal resistance in the circuit.

Similarly, since the 1990s, there have been several studies showing spectacular improvement (and in many cases, degradation) in physical properties of nanofluids. Often, these studies have involved measurement of (or predictions for) thermal characteristics as well as rheological properties and convective behavior of nanofluids. In contrast, for the nanocomposites—the enhancements in similar material characteristics have not lived up to the expectations for the values predicted from theoretical analyses. Nanofluids are also plagued by the precipitation of NP—an issue that has not received its due respect or consideration in the thermofluidics literature. The precipitation of NP from the nanofluid effectively results in the formation of nanocoatings on the heat exchanging surface. Hence, initially, it was not clear if the convective heat transfer enhancements (or degradation) reported in the thermofluids literature for nanofluids is dominated by the material properties of the nanofluid or if it is due to modification of the interfacial effects on the heat exchanging surface arising from the precipitation of the NP. The nanoparticle precipitation can progressively build up over time to form layers of nanocoatings leading to fouling of the heat exchanging surface and therefore degradation in the resultant heat transfer over time while conducting the experiments. In contrast, if the level of precipitation is limited to isolated NP on the heat exchanging surface—it can lead to enhancement of the effective surface area and therefore potentially in the resultant heat transfer, especially if the duration for conducting the experiments is short enough. The enhancement of the surface area in a heat exchanger by such engineered surface structures can be termed as "extended surface" or "fins" (i.e., nanofins). Hence, nanofins were shown to form during flow of nanofluids on heater surfaces (Yu 2012). Further experiments (Yu 2012) showed that repeating the control experiment—i.e., flowing pure solvent (fluid without NP) once again—after conducting the nanofluid experiments using the same apparatus resulted in similar levels of heat transfer enhancement with the pure solvent. In other words, the sequence of conducting the control experiments resulted in different levels of heat transfer for the same solvent. Microscopy of the heater surfaces showed that the heater surfaces were covered with NP that had precipitated from the nanofluids—leading to formation of nanofins. To further prove this nanofin hypothesis, confocal microscopy experiments were performed to measure the wall temperature gradient using water with fluorescent dyes and aqueous nanofluids (water with quantum dots, QDs). The results showed that nanofluid samples demonstrated 3–4 times higher wall temperature gradient than the pure coolant (with dye). In a further proof of the nanofin effect—experiments were conducted using silicon nanofins—which showed same level of enhancement when compared to control experiments conducted using plain heater surfaces (without nanofins) as the nanofluid experiments performed on plain heater surfaces (without nanofins). These experiments (Yu 2012) demonstrated that the heat transfer enhancements observed for nanofluids are also dominated by the nanofin effect and the fluid properties play a secondary role.

Motivation Statement

Advances in microelectronics technology have led to miniaturization of electronic devices with the added advantage of enhanced performance of each chip. This advancement is also associated with enhanced heat flux from these chips leading to increase in demand for higher cooling requirements. Effective thermal management of these devices has become a major technological challenge for the microelectronics industry. Air-based cooling systems are more common and reliable and are still favored for cooling of low heat flux electronic devices. Attempts have been made to improve the thermal transport of these systems using fins and high-performance fans, but due to the poor thermal characteristics of air, the effectiveness of these systems is compromised. Therefore, in many high heat flux applications liquid cooling is preferred. Liquid cooling is beneficial due to the superior thermal properties of liquids. Liquid cooling can be either single-phase flow or multi-phase flow with phase change. Phase change enables superior cooling performance when compared to the single-phase cooling due to the high latent heat. Other candidate applications include:

- Biotechnology, Homeland/Biosecurity: Nanofin structures can be used to enhance the device performance that requires thermocycling (e.g., for genomic signal amplification and diagnostics using polymerase chain reaction that involves repeated and rapid thermocycling).
- Deep drilling for oil and gas exploration (>15,000 ft): Cooling of electronics under high temperature and pressure conditions are needed (e.g., DOE Deep-Trek program). Nanofins and robust nanosensors (e.g., thin-film thermocouples, TFT) can be used to mitigate the cooling requirements in such harsh environments.
- Energy Efficiency/Sustainability: Incorporation of nanofins can enable the development of more efficient heat exchangers, which are useful for energy efficient buildings and heating, ventilation, and air conditioning (HVAC) systems.

Similarly, nanofluids have been reported for enhanced thermophysical properties such as thermal conductivity and specific heat capacity. Also, nanofluids may display non-Newtonian rheology, while the pure solvent itself maybe Newtonian. This implies that certain composition of nanofluids may be better suited as heat transfer fluids (HTF) owing to the enhanced thermal conductivity, while certain other formulations may enable the nanofluids to be more amenable as thermal energy storage (TES) materials owing to their enhanced specific heat capacity. A candidate application is thermal power generation technologies, such as for TES devices used in improving the energy efficiency of concentrated solar power (CSP).

Nanofins and Nanocoatings: A Historical Perspective

In cooling applications, pool boiling is simple to achieve, reliable, quiet, and inexpensive. According to Mudawar and Anderson (1990), there are three ways to enhance heat transfer in pool boiling:

1. modifying the boiling surface,
2. subcooling the liquid, and
3. increasing the operating pressure.

Subcooling of liquid requires extra equipment and increasing operating pressure for thermal management of electronic devices is often not a viable option. Hence, modifying the boiling surface is the most cost-effective approach for enhanced cooling in commercial endeavors.

Surfaces can be modified by texturing (e.g., chemical etching) or coating. Boiling experiments on nanotextured surfaces have shown significant increase in the heat transfer. Pool boiling experiments are performed on silicon substrates coated with CNT and etched silicon nanofins. Experiments performed on silicon heaters coated with CNT have shown 60 % increase in the critical heat flux (CHF) as compared to the bare silicon surfaces in pool boiling (Ahn et al. 2006; Sathyamurthi et al. 2009; Ujereh et al. 2007). Also, nanotube coating with thickness of 25 μm was found to increase the film boiling heat flux by 60–150 % as compared to the bare silicon surface. However, nanotube coatings with thickness of 9 μm failed to show any enhancement in film boiling (Sathyamurthi et al. 2009).

Pool (nucleate) boiling experiments performed by Sriraman et al. (Sriraman 2007) on silicon substrates with etched silicon nanofins showed heat flux enhancement of as much as 120 % as compared to the bare silicon surface in nucleate boiling. Also, in nucleate boiling, the enhancement in heat flux was found to be independent of the length of the nanotube. To explain this phenomenon, consider a nucleation site and a vapor bubble attached on the nanofins fabricated on the silicon surface as shown in Fig. 1.1.

According to Derjaguin and Zorin (1957), there is a very small layer of liquid film that adheres on the silicon surface (under the vapor bubble) which is termed as the "microlayer." This microlayer acts as a barrier to the heat flow from the heater surface to the vapor surface. The height of this microlayer is estimated in the literature to be of the order of 100 nm (Sriraman 2007). Nanofins of length greater than 100 nm disrupt the microlayer and protrude inside the vapor bubble as shown in Fig. 1.1. This leads to the disruption of the microlayer and increases area of liquid microlayer in contact silicon surface. Hence, the heat transfer from the silicon surface increases. The length of nanofin inside the vapor bubble is not in contact with the liquid and experiences insulation. So, nanofins of height greater than 100 nm are equally effective as nanofins of length of 100 nm and therefore do not enhance heat transfer any further when the length is extended beyond 100 nm.

In case of film boiling, a layer of vapor is formed on the silicon heater that separates the liquid phase from the heater substrate. The vapor film serves as an insulator which can periodically break down leading to transient liquid–solid contact and very high levels of transient heat conduction which are ephemeral. The minimum thickness of this vapor film is of the order of 15–20 μm (Banerjee and

Fig. 1.1 Sketch of the (*top*) bubble on a solid surface with nanofins during pool boiling showing the presence of thin-film liquid layer (microlayer) under the bubble (*bottom*) perturbation and disruption of the microlayer due to the presence of the nanofins

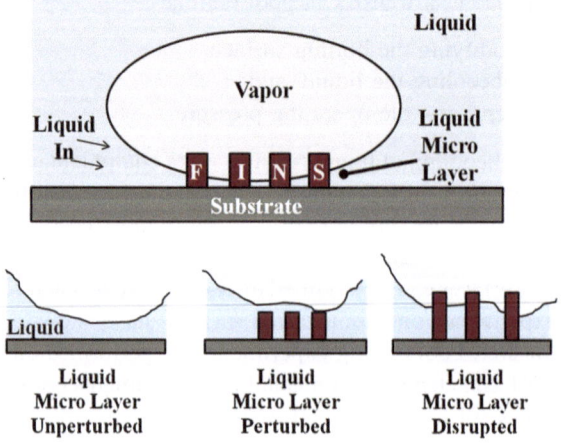

Dhir 2001; Banerjee et al. 1996). NT with height greater than 15–20 μm disrupt the vapor layer and protrude into the liquid, thus forming an effective bridge for conducting heat from the substrate to the liquid phase and bypassing the insulating vapor film—therefore leading to greater efficacy in enhancing the heat transfer. This mechanism of disruption of vapor film by long NT is depicted in Fig. 1.2.

Experiments have shown that NT of length of 25 μm enhance heat transfer, while NT of length of ~ 10 μm do not show any enhancement as compared to that on a bare silicon wafer as the heater substrate (Sathyamurthi et al. 2009).

From the discussions above, it can be seen that both silicon nanofins and CNT are able to enhance pool boiling heat flux. The heat flux augmentation was attributed to combined effect (i.e., "nanofin effect") arising from the increase in the surface area due to the protruding NT in combination with the augmentation of thermal resistance in the form of Kapitza resistance, disruption of vapor films, and modification of the thermal/mass diffusion boundary layers (since the nanofins or NT also induce local chemical concentration gradients of the solvent molecules/species in their vicinity).

Several studies (e.g., Singh 2010; Unnikrishnan et al. 2008; Singh et al. 2011; Jo and Banerjee 2011a) identified the thermal interfacial resistance (or "Kapitza resistance") as the dominant parameter for heat flux enhancement that were observed for various nanofins.

In addition to the experiments mentioned above, several other experimental results exist in the literature involving nanocoatings and nanostructured surfaces involving inorganic materials. However, these reports in the literature suffer from a serious drawback—arising from a flaw in the design of experiments. These experiments were performed using heaters that are smaller in size than the "infinite heater limit." In the infinite heater limit, the boiling heat flux values are independent of the size of the heater. Thus, if the heater size is smaller than this limit (i.e., if it is a "small heater")—the boiling heat flux is a function of the size of the heater. In addition, porous layers (such as nanocoatings) can cause reduction in the size of the instability wavelengths at the vapor–liquid interface. Hence, the same heater can be a small heater without nanocoatings and can display infinite heater behavior on application of nanocoatings. Hence, it is recommended that the pool boiling experiments be conducted on heaters with length larger than 5 cm for polar or ionic liquids (such as water) and length larger than 2 cm for nonpolar liquids (e.g., organic refrigerants). Further details on this discussion are available in the book chapter by Singh et al. (2012). However, for the completeness of the

Fig. 1.2 Vapor blanket on the heater surface in film boiling with nanotubes of different lengths

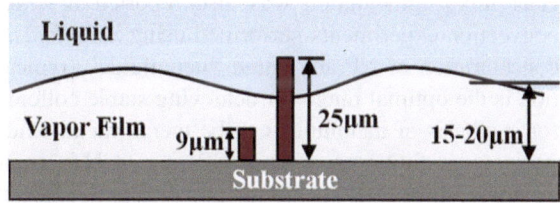

discussions, these reports in the literature are briefly discussed here with the caveat that serious logical conclusions are not amenable from these studies owing to the serious flaws in the design of experiments.

Chen et al. (2009) reported that pool boiling CHF of water was enhanced by ~ 100 % for surfaces with copper and silicon nanowires. Wu et al. (2010) reported that CHF of water and FC-72 was enhanced by ~ 50 and ~ 40 % for substrates with TiO_2 nanocoatings. Hendricks et al. (2010) reported that CHF of water was enhanced by ~ 400 % for Al and Cu substrates with ZnO nanocoatings. Im et al. (2010) reported that CHF of water on silicon substrate with Cu NW was enhanced only marginally, while the heat transfer coefficient was enhanced by a factor of ~ 10. Thus, the results of Im et al. (2010) are in direct contradiction to the results reported by Chen et al. (2009)—primarily due to flawed design of experiments in both studies. Also, Launay et al. (2006) reported that CHF of water and PF-5060 was enhanced by ~ 60–900 % for substrates with CNT nanocoatings, which is also inconsistent with other reports in the literature owing to faulty design of experiments as well as in the experimental procedure (e.g., criterion for establishment of steady-state conditions).

Flow boiling experiments were reported by Sunder and Banerjee (2009) using nanothermocouples where the thin films deposited by physical vapor deposition (PVD) techniques were ~ 200 nm thick. Subsequently, Singh et al. (2010) reported that flow boiling heat transfer was enhanced by as much as 180 % on silicon substrates coated with CNT. Khanikar et al. (2009a, b) reported that significant enhancement of flow boiling heat fluxes were observed at low mass velocities—while Khanikar et al. (2009a, b) showed that there was marginal or no enhancement at higher mass velocities. This is consistent with the trends reported by Singh et al. (2010)—since the proportion of the phase change heat flux is weakly sensitive to flow rate and remains almost constant. Hence, the proportion of phase change heat flux in the total heat flux decreases at higher flow rates as the convective heat transfer starts dominating the proportion of the total heat flux.

In summary, these boiling experiments involving organic nanocoatings (CNT), inorganic nanocoatings (MNW), and silicon nanofins show that the nanofin effect—primarily the Kapitza resistance—is the dominant factor that dictates the peculiar behavior observed in these experiments.

Nanofins and Nanofluids: A Historical Perspective

Heat flux augmentation was also reported in several studies involving forced convection experiments performed using *nanofluids*. Coolants doped with a small concentration of NP are called "nanofluids" (typically 0.1–1 % mass concentration is the optimal range for achieving stable colloidal suspensions). The NP used for synthesis of nanofluids can be metal, metal oxide/ceramics (e.g., TiO_2), inorganic (e.g., SiO_2) or organic NP (e.g., CNT, graphene), which form colloidal mixtures, since their size range is typically 1–100 nm. Heat flux was enhanced by

10–30 % for poly-alpha-olefin (PAO) coolant doped with 0.6 % mass concentration of organic NP (Ding et al. 2006; Huitink et al. 2007; Jackson et al. 2006; Kakac and Pramuanjaroenkij 2009; Nelson et al. 2009). Nelson et al. (2009) reported that these NP precipitate on the heater surface, effectively forming *nanofins*.

Several studies have been reported in the literature to elucidate the perceived anomalous enhancement of thermal conductivity and viscosity of nanofluids. In contrast, Boungiorno et al. (2009) reported a benchmarking study by multiple investigators for thermal conductivity measurements for various nanofluid samples (aqueous nanofluid and olefin nanofluid samples) using different measurement techniques—which are primarily variants of the hot-wire technique (and all of the measurements involve a measurement probe in physical contact with the nanofluid samples). The measurements were performed by multiple investigators in different research laboratories using "blind" samples. The authors concluded after compiling the experimental results from various investigators participating in this study that the thermal conductivity values were consistent with the Maxwell's model and the authors concluded that the anomalous enhancement reported in prior studies was not observed for the nanofluids considered in that study. It is not clear if the conclusions from this study (Boungiorno et al. 2009) can be applied to all classes of nanofluids or if this study was relevant and valid for only the nanofluids used in that study. Also, it may be noted that data sets that proved to be outliers were discarded from the compiled data sets—since they were obtained from multiple investigators participating in this study (while the discarded data sets were not reported in this paper). Discarding experimental data sets that are outliers from an expected trend—is an erroneous and questionable strategy (at the least, these discarded data sets should have been reported in an Appendix or as Supplementary Materials). Such decisions have been known to plague famous researchers and have been criticized in posterity (this temptation seems to have plagued even Nobel Prize winners such as Linus Pauling).

In contrast to the vast number of studies in the literature on thermal conductivity of nanofluids, only a few studies have been performed to measure the specific heat capacity of nanofluids. Namburu et al. (2007) reported that the specific heat capacity of aqueous SiO_2 nanofluids decreased with increase in the nanoparticle concentration. Similarly, Zhou and Ni (2008) reported the same trend of decreasing specific heat capacity for aqueous dispersion of Al_2O_3 NP. Vajjha and Das (2009) reported peculiar trends in the specific heat capacity measurements of performed the measurement of water-/ethylene glycol-based nanofluids using NP of Al_2O_3, ZnO, and SiO_2. They reported degradation in the specific heat capacity values below a critical threshold temperature of the nanofluid samples and marginal enhancement when the temperature of the sample exceeded the threshold temperature of ~ 70 °C compared with that of the neat solvent. Zhou and collaborators reported similar results involving degradation of the specific heat capacity for ethylene glycol–CuO nanofluids (Zhou et al. 2009).

Interestingly, studies involving nonaqueous solvents showed radical enhancements in the specific heat capacity. Nelson et al. reported that the specific heat capacity of a PAO nanofluid mixed with graphite NP (graphene sheets) (Nelson et al. 2009) was enhanced by 50 % for nanoparticle mass fraction of 0.6 %. Shin and Banerjee (2010, 2011a, b, c, 2013) as well as Shin (2011) reported the enhancement in the specific heat capacity by as much as 20–120 % on doping with silica NP in molten salt eutectics (Jo and Banerjee 2011b; Jung and Banerjee 2011; Shin and Banerjee 2011a, b, c, 2013). Enhancements in the specific heat capacity of molten salt-based carbon nanotubes nanofluids were reported for both solid and liquid phase (Jo and Banerjee 2010). Similar results were also reported by Jung (2012) and Jo (2012).

Bridges et al. (2011) reported that the volumetric heat capacity of an ionic liquid was enhanced by ~45 % on doping with alumina NP. These samples were termed as "NEIL" or "nanoparticle-enhanced ionic liquids."

It is interesting to note that all of the studies in which the decrease in the specific heat capacity were reported—used conventional fluids such as water or EG as the solvent medium for nanofluids. On the other hand, the studies that reported the enhancement of the specific heat capacity (or volumetric heat capacity) values employed different solvent materials, such as carbonate molten salt eutectic, ionic liquids, or PAO. Here, the material properties of the solvent are expected to be the dominant parameter rather than that of the NP.

Results reported by Singh (2010), Shin (2011), Jung (2012), and Jo (2012) showed that for the MD simulations involving molten salt suspension of NP (Jo and Banerjee 2011a)—the NP induce the formation of a "compressed phase" of solvent molecules on the surface of the NP. The density of the compressed phase was predicted to match the solid phase density of the solvent material. In addition, the MD simulations showed that the compressed phase formed due to adsorption of solvent molecules on the nanoparticle surface can in turn induce a concentration gradient of the molecules (for a mixture of different materials or species constituting the solvent phase).

According to Jo (2012): "The nanoparticle is likely to have higher physical/chemical affinity (arising from van der Waals interactions, ionic interactions, and the combination thereof) for a particular species ('preferred species') in the solvent mixture—causing a higher concentration of the preferred species on the nanoparticle surface—compared to in the bulk. The concentration of the preferred species on the nanoparticle surface was also predicted to vary with the changes in the composition of the bulk phase of the solvent (or neat solvent, i.e., without the nanoparticle present). The changes in concentration of the preferred species on the nanoparticle surface are also likely to cause minor perturbations in the density of the resultant compressed phase that forms on the surface of the nanoparticle. Such a complex interaction is also likely to affect the material properties and transport properties on the molecular scale for the nanoparticle suspension. This is also likely to cause spatial variation in material properties in an around (in the vicinity) of the NP in the mixture."

According to Shin (2012) and Shin and Banerjee (2011a), "Based on MD simulations reported in the literature, the density of the semisolid compressed layer was assumed to be the same as the matrix/solid phase property (of the solvent material). Also, based on the MD simulations, the property values were assumed to be at the melting point of the pure solvent material."

Hence, the nanofin effect is also responsible for the changes in the specific heat capacity (as well as density, viscosity, and thermal conductivity) of nanofluids. In other words, the introduction of a nanoparticle (or nanofin) induces effects similar to phase change in the vicinity of the nanofin and thereby affects bulk the property values of the mixture or the fluid in contact with the substrate with surface nanostructures.

Experimental measurements using nanothermocouple array (also called "TFT") were reported by Yu (2012) for flow of aqueous nanofluids (i.e., using de-ionized water/DIW as the neat solvent) in heated microchannels show that the nanofins formed by precipitation of NP is the dominant mechanism for heat transfer enhancements—while the properties of the coolant fluid (i.e., DIW or nanofluid) play a secondary (or marginal) role. As reported by Yu (2012): "Initial experiments were performed using a microchannel apparatus integrated with TFT array that was fabricated in situ. This experimental apparatus was then mounted on the stage of a confocal microscope for performing flow visualization experiments. The flow visualization experiments were performed using: (1) laser-induced fluorescence (LIF) techniques using organic dye, fluorecein, and (2) quantum dots (QDs). In addition, the level of precipitation of the NP was monitored after each experiment using materials characterization techniques such as SEM and EDX.

Furthermore, different strategies were implemented in the experimental procedure. The order of performance of the experiments was found to affect the heat flux values. For example, a control experiment is performed by flowing DIW in the heated microchannel. This was followed by experiments performed using nanofluid coolants (SiO_2 nanofluids or TiO_2 nanofluids). Subsequently, the control experiment is repeated by flowing DIW in the same heated microchannel. It was observed that repeating the control experiment after the nanofluid experiments, consistently yielded heat flux values that matched (or in certain cases exceeded) that of the nanofluids experiments and almost always exceeded the heated flux values obtained from the initial control experiments.

Also, artificial nanofins are fabricated using SFIL technique. The height of nanofins can be easily controlled by varying DRIE cycles during the fabrication processes. TFT arrays are fabricated on nanofin surface, and the experiments were performed using DIW as a test fluid. These experiments showed that the heat transfer coefficient of DIW on nanofin surface is higher than that of DIW on plain surface, which supports the hypothesis (nanofin effect) of the current study.

Hence, these experiments were used to prove the hypothesis central to this study: that the isolated precipitation of the NP leads to formation of nanofins on the heated surface that enhances the effective surface area for forced convective heat transfer in microchannels, and this is the most dominant transport mechanism of nanofluids flowing in the microchannel. That is, thermophysical properties of the nanofluid coolants plays minor role in transport mechanism compared to the surface modification. In addition, the implication of this hypothesis was also proven that excessive agglomeration and precipitation of NP lead to the formation of a fouling film that acts as an additional thermal barrier/resistance to heat transfer due to scaling (fouling)—thus causing degradation of the forced convective heat transfer (compared to that of the control experiments performed using DIW coolant).

Several anomalous (or counterintuitive) behaviors of the nanofluid coolants were also observed in these experiments. These are summarized as follows:

1. SiO_2 nanofluids were observed to cause either enhancement or degradation of the heat flux values, depending on the level of precipitation (and surface fouling). In contrast, TiO_2 nanofluids were consistently observed to cause enhancement in heat flux (compared to the control experiments performed using DIW).
2. The level of precipitation for TiO_2 nanofluids was observed to be much less compared to that of the SiO_2 nanofluids in the SEM images (possibly due to better stability of the nanofluids and lower affinity of the NP for the Pyrex wafer substrate).
3. TFT experiments were observed to yield lower values of heat flux than that of the flow visualization experiments (possibly due to nonuniform heat flux distribution, localized heating leading potentially to phase change/formation of nanobubbles due to laser illumination, higher local values of heat flux obtained by point measurements rather than the lower values of global heat flux measured due to line-average measurements, etc.).
4. The nanofluids experiments seem to render "system memory" to the experimental apparatus—due to formation of nanoparticle precipitates. Thus, when the control experiment is repeated before and after the nanofluids experiments—consistently higher values of heat flux are obtained that usually matches (or exceeds) the heat flux values obtained from the nanofluids experiments. This proves the hypothesis central to this study (i.e., nanofin effect dominates over the thermophysical property effects and is the primary driver in the heat flux enhancements that are observed in nanofluids experiments).
5. Artificial nanofins are successfully fabricated using SFIL process. TFT is integrated with these nanofins and convective heat transfer experiments were performed on this apparatus using DIW as a testing fluid. The results showed higher heat flux values compared to those on the plain surface using the same fluid. The level of enhancement is quite similar to that of using nanofluids. These results add credence to the nanofin hypothesis that has been proposed and validated in this study.

It is revealed that the precipitation of NP is the most dominant heat transfer mechanism of nanofluids. Isolated precipitation of NP behaves as nanofins, which results in significant enhancement of heat transfer coefficient. However, excessive precipitation forms fouling film and causes the degradation of heat transfer. Artificial nanofins are fabricated using SFIL process to evaluate the significance of precipitation effect. When these nanofins are tested in the experiment, similar level of enhancement in forced convection heat transfer coefficient has been observed compared to those of nanofluids. This result demonstrates and validates nanofin hypothesis that the precipitation of NP is predominant mechanism of heat transfer enhancement using nanofluids.

Considering the experimental results of the current study, there are several benefits to utilize nanofins and microchannel with respect to the effective method for the enhancement of heat transfer. First of all, a microchannel has an advantage over a macrochannel in that a former provides much larger surface area with the same volume. Therefore, the level of enhancement of heat transfer in micro-channel by introducing either nanofins or isolated precipitation of NP would be greater than that of in macrochannel. Using nanofins are much reliable than using nanofluids if the technique is adopted in the applications because repeated use of nanofluids possibly leads excessive precipitation of NP and degradation of heat transfer. In addition, nanofluids cause increase in pumping cost since the viscosity of nanofluids is higher than pure solvent, which does not have to be worried about when nanofins are used. However, fabrication of nanofins is expensive compared to just using nanofluids".

Thus, the results reported by Yu (2012) demonstrate that nanofluids can serve as an efficient vehicle (or means) for realizing nanocoatings on heater surfaces and flow conduits. These nanocoatings in turn behave as a disordered array of nanofins and are once again subject to the nanofin effect. Hence, subliminally, the phenomenon responsible for heat transfer enhancement in boiling and nanofluids is the same—which is the "nanofin effect." This is discussed next.

Moral of the Story

Nanofins formed on the heater surfaces—from nanocoatings, etched surface nanostructures, or NP precipitated from flowing nanofluids—typically lead to enhancement in heat transfer. The factors responsible for the enhancement in heat transfer are either due to the high thermal conductivity of the nanofin material or due to the enhancement in the surface effects ("nanofin effect"). The thermal conductivity of silicon is of the order of 150 W/mK (Shanks et al. 1963), whereas for CNT, it is of the order of 3,000 W/mK for multi-walled NT (Kim et al. 2001) and 6,000 W/mK for single-wall CNT (Berber et al. 2000). Since the enhancement in heat transfer is observed experimentally in case of both silicon and carbon nanofins, therefore rather than thermal conductivity, surface effect is the more dominant factor. The primary factor limiting the enhancement in heat transfer performance of carbon nanofins, even though there is enhancement in surface area, is the presence of interfacial resistance to the heat flow from the surface of the CNT or nanofins to the surrounding fluid.

Hence, the objective of this book is to explore and demonstrate the various factors affecting the interfacial thermal resistance between a nanofin and the surrounding fluids. In essence, it will be shown that the effect of molecular composition and molecular structure of the surrounding fluids affects interfacial thermal resistance. In addition, a recent study Hu and Banerjee (2013) has shown that the chirality of a CNT can also affect the interfacial resistance between a nanofin and the surrounding fluid molecules. For example, metallic CNT are estimated to have different interfacial thermal resistance than semiconducting CNT.

Various Aspects of the Nanofin Effect

Chemical composition of a coolant can play a vital role in the heat transfer from the CNT. The differences in the atomic structure of the molecules with the same type and number of atoms can lead to very different properties of the system. In order to study the performance of the NT as nanofins in different types of fluids (HTF or TES), characterization of the effect of the atomic (chemical) and structural properties of the molecules on the interfacial thermal resistance needs to be studied systematically. Interfacial resistance in nanostructures dominates the overall thermal resistance and significantly affects the thermal transport properties. The main focus of this study will be on using nonequilibrium molecular dynamics simulations to explain the effect of chemical and structural properties on the interfacial resistance between coolant molecules and a carbon nanotube acting as *nanofin*.

Common coolants used for heat removal such as water, ethyl alcohol, 1-hexene and single long-chain hydrocarbons, n-heptane and its isomers, chains and their mixtures, manufactured by the catalytic oligomerization of poly-alpha-olefins (PAOs) are considered here. PAOs are commonly used as TES and HTF, such as for cooling electronics platforms, especially for avionics cooling applications— owing to their superior physical and chemical properties, such as greater fluidity at low temperature, lower volatility, a higher viscosity index, lower pour point, better oxidative and thermal stability, and low toxicity (Benda et al. 1996). PAOs are manufactured via catalytic oligomerization of alpha-olefins. For avionics applications, a second catalytic process breaks the double bonds in PAOs, producing a mixture of chemically saturated polymers. These polymers possess much lower pour point suitable for cooling electronics platforms.

Significance of the Nanofin Effect

Recent literature reports explored the use of carbon nanotube arrays as nanofins for cooling applications (Kordas et al. 2007; Sriraman 2007; Ujereh et al. 2007). The major hurdle to the heat dissipation by these nanofins is due to the dominance of the interfacial thermal resistance between the nanofins and the surrounding fluid. To better understand the performance of these nanofins, comprehensive synthesis of information about the factors affecting interfacial thermal resistance is of paramount importance. At the molecular scale, chemical and structural properties of the molecules play a vital role in determining the system behavior. This is the first study to systematically elucidate the effect of atomic and structural variations on the interfacial thermal resistance between the CNT and coolants. The few studies performed so far only focused on calculating the magnitude of the interfacial thermal resistance between a carbon nanotube and surrounding fluid molecules. The significant aspects of the nanofin effect that are explored in this book include:

- The effect of chemical composition of fluids on the interfacial thermal resistance.
- The effect of molecular structure of the fluids on the interfacial thermal resistance.
- The effect of polymer chains of a molecule on the interfacial resistance.
- The effect of isomers of a polymer chain on the interfacial thermal resistance.
- The effect of mixture of polymer chains and their isomers on the thermal resistance.
- Dominant mechanism of energy transfer from CNT to surrounding molecules.

Role of Material Properties in the Nanofin Effect

Experiments have shown enhancement in heat transfer using both carbon and silicon nanofins. The factors responsible for enhancement in the heat transfer are increase in the surface area due to the presence of the nanofins leading to disruption of vapor films and modification of the thermal/mass diffusion boundary layers and high thermal conductivity of the nanofins. CNT have much higher thermal conductivity than silicon nanofins, but experiments have shown enhancement in heat transfer using both types of nanofins. So, enhancement in surface area is the dominant factor in enhancing the heat transfer. Figure 1.3 shows the thermal resistance model for a single nanofin.

From the figure, the following thermal resistances are identified, the resistance due to the finite thermal conductivity of the silicon substrate R_1, resistance at the nanofin-silicon substrate interface R_2, resistance due to the finite thermal conductivity along the length of the nanofin R_3 and the resistance to the heat transfer from the nanofin to the surrounding liquid molecules R_k also known as "interfacial thermal resistance" or "Kaptiza resistance." The resistance R_1 is not as important in this context since it is the heater surface that is common to both configurations (i.e., for a flat substrate and a substrate with nanofins). The resistance R_2 is only present in case of carbon nanofin (i.e., CNT) and is of the order of $\sim 10^{-10}$ m^2 K/W (Maune et al. 2006), and absent for the silicon nanofins as they are etched on the silicon surface (heater)—so there is no contact resistance. The resistance R_3 is very small owing to the small length of nanofins (of the order of $\sim 10^{-13}$ m^2 K/W). Interfacial thermal resistance, R_4, is of the order of $\sim 10^{-11}$ m^2 K/W (Murad and Puri 2008) in case of silicon nanofins,

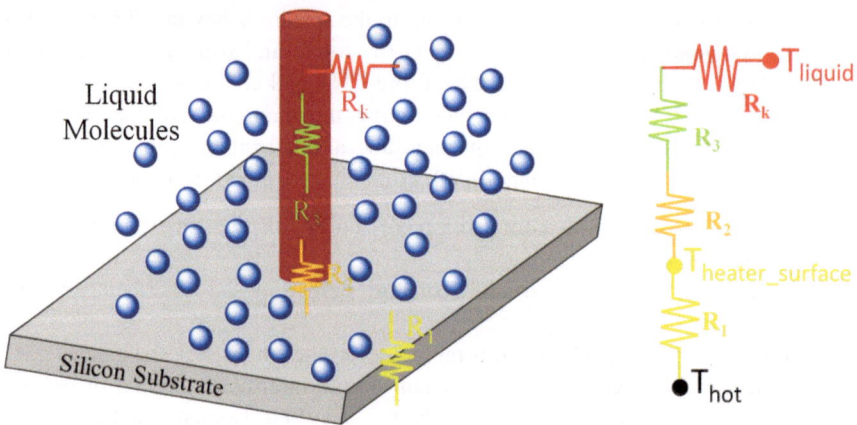

Fig. 1.3 Cartoon with resistance model for carbon nanofin on a silicon substrate

while it is the order of $\sim 10^{-8}$ m^2 K/W (Huxtable et al. 2003) in case of carbon nanofins (CNT). So, in case of CNT, the interfacial resistance is three orders of magnitude higher than that of silicon nanofins. This shows that rather than conductive resistance, the interfacial resistance is the more dominant factor in modulating heat flux from nanostructures. Hence, this resistance (R_k) dominates the total resistance in the thermal impedance network and severely limits the effectiveness of the carbon nanofins (in comparison with that of the silicon nanofins).

NT have also been used as a filler material in composites to increase thermal conductivity. NT are expected to enhance the thermal conductivity of the composites many fold owing to their high aspect ratio and very high thermal conductivity. But experiments show small increase in the thermal conductivity (Biercuk et al. 2002; Bryning et al. 2005; Choi et al. 2001) compared to the theoretical calculations. This anomaly has been attributed to the interfacial thermal resistance between the CNT and the polymer composite molecules.

The discussions in this book are limited to the investigation of a (5,5) carbon nanotube as a nanofin. Also, for the matrix system (neat solvent): water, alcohol, 1-hexene, n-heptane and its dimers, trimers, and isomers are studied. The isomers of these polymers considered in this study are limited to a single methyl group being placed at different locations on the polymer chain.

As mentioned earlier, a recent study Hu and Banerjee (2013) has shown that the chirality of a CNT can also affect the interfacial resistance between a nanofin and the surrounding fluid molecules. For example, metallic CNT are estimated to have lower interfacial thermal resistance than semiconducting CNT.

In summary, nanofin effect is ubiquitous, especially as it relates to application of nanotechnology in thermofluidics. The nanofin effect arises from the dominance of Kapitza resistance (R_k) in the thermal impedance network. Typically, noncontinuum models are used to estimate R_k. Molecular simulation techniques provide amenable tools for obtaining these estimates. This is discussed in the next chapter.

References

Ahn HS, Sinha N, Zhang M, Banerjee D, Fang S, Baughman RH (2006) Pool boiling experiments on multiwalled carbon nanotube (mwcnt) forests. J Heat Trans 128:1335–1342

Banerjee D, Dhir VK (2001) Study of subcooled film boiling on a horizontal disc: part 2—experiments. J Heat Trans 123:285–293

Banerjee D, Son G, Dhir VK (1996) Conjugate thermal and hydrodynamic analyses of saturated film boiling from a horizontal surface. ASME Heat Transfer Div Publ HTD 334:57–64

Benda R, Bullen J, Plomer A (1996) Synthetics basics: polyalphaolefins-base fluids for high-performance lubricants. J Synth Lubr 13:41–57

Berber S, Kwon YK, Tomanek D (2000) Unusually high thermal conductivity of carbon nanotubes. Appl Phys Lett 84:4613–4616

Biercuk MJ, Llaguno MC, Radosavljevic M, Hyun JK, Johnson AT, Fischer JE (2002) Carbon nanotube composites for thermal management. Appl Phys Lett 80:2767–2769

Bridges NJ, Visser AE, Fox EB (2011) Potential of nanoparticle-enhanced ionic liquids (neils) as advanced heat-transfer fluids. Energy Fuels 25:4862

Bryning MB, Milkie DE, Islam MF, Kikkawa JM, Yodh AG (2005) Thermal conductivity and interfacial resistance in single-wall carbon nanotube epoxy composites. Appl Phys Lett 87:161909

Buongiorno J, Venerus DC, Prabhat N, McKrell T, Townsend J, Christianson R, Tolmachev YV, Keblinski P, Hu L-W, Alvarado JL (2009) A benchmark study on the thermal conductivity of nanofluids. J Appl Phys 106:094312–094312-094314

Chen R, Lu M-C, Srinivasan V, Wang Z, Cho HH, Majumdar A (2009) Nanowires for enhanced boiling heat transfer. Nano Lett 9:548–553

Choi SUS, Zhang ZG, Yu W, Lockwood FE, Grulke EA (2001) Anomalous thermal conductivity enhancement in nanotube suspensions. Appl Phys Lett 79:2252–2254

Derjaguin BV, Zorin ZM (1957) Optical study of the absorption and surface condensation of vapors in the vicinity of saturation on a smooth surface, vol 2. In: Proceeding of 2nd international cogress on surface activity, London, pp 145–152

Ding Y, Alias H, Wen D, Williams RA (2006) Heat transfer of aqueous suspensions of carbon nanotubes (cnt nanofluids). Int J Heat Mass Transf 49:240–250

Hendricks TJ, Krishnan S, Choi C, C-H Chang, Paul B (2010) Enhancement of pool-boiling heat transfer using nanostructured surfaces on aluminum and copper. Int J Heat Mass Transf 53:3357–3365

Hu Y, Banerjee D (2013) Numerical investigation of the effect of chirality of carbon nanotube on the interfacial thermal resistance. J Nanofluid 2:29–37

Huitink D, Ganguly S, Banerjee D, Yerkes K (2007) Convective heat transfer enhancements using nanofluids. In: Proceedings of the nanofluids: fundamentals and applications (engineering conferences international)

Huxtable ST, Cahill DG, Shenogin S, Xue L, Ozisik R, Barone P, Usrey M, Strano MS, Siddons G, Shim M (2003) Interfacial heat flow in carbon nanotube suspensions. Nat Mater 2:731–734

Im Y, Joshi Y, Dietz C, Lee SS (2010) Enhanced boiling of a dielectric liquid on copper nanowire surfaces. Int J Micro-Nano Scale Transp 1:79–96

Jackson JE, Borgmeyer BV, Wilson CA, Cheng P, Bryan JE (2006) Characteristics of nucleate boiling with gold nanoparticles in water. In: Proceedings of the IMECE 2006, IMECE2006-16020

Jo B (2012) Numerical and experimental investigation of organic nanomaterials for thermal energy storage and for concentrating solar power applications. Dissertation, Texas A&M University, College Station

Jo B, Banerjee D (2010) Study of high temperature nanofluids using carbon nanotubes (cnt) for solar thermal storage applications. ASME

Jo B, Banerjee D (2011a) Interfacial thermal resistance between a carbon nanoparticle and molten salt eutectic: effect of material properties, particle shapes and sizes. In: ASME/JSME 8th thermal engineering joint conference, Honolulu, HI, pp 13–17

Jo B, Banerjee D (2011b) Enhanced viscosity of aqueous silica nanofluids. Dev Strateg Mater Comput Des II: Ceram Eng Sci Proc 32:139–146

Jung S (2012) Numerical and experimental investigation of inorganic nanomaterials for thermal energy storage (tes) and concentrated solar power (csp) applications. Dissertation, Texas A&M University, College Station

Jung S, Banerjee D (2011) Enhancement of heat capacity of nitrate salts using mica nanoparticles. Dev Strateg Mater Comput Des II: Ceram Eng Sci Proc 32:127–137

Kakac S, Pramuanjaroenkij A (2009) Review of convective heat transfer enhancement with nanofluids. Int J Heat Mass Transf 52:3187–3196

Khanikar V, Mudawar I, Fisher T (2009a) Effects of carbon nanotube coating on flow boiling in a micro-channel. Int J Heat Mass Transf 52:3805–3817

Khanikar V, Mudawar I, Fisher TS (2009b) Flow boiling in a micro-channel coated with carbon nanotubes. IEEE Trans Compon Packag Technol 32:639–649

Kim P, Shi L, Majumdar A, McEuen PL (2001) Thermal transport measurements of individual multiwalled nanotubes. Appl Phys Lett 87:215502

Kordas K, Toth G, Moilanen P, Kumpumaki M, Vahakangas J, Uusimaki A, Vajtai R, Ajayan P (2007) Chip cooling with integrated carbon nanotube microfin architectures. Appl Phys Lett 90:123105

Launay S, Fedorov A, Joshi Y, Cao A, Ajayan P (2006) Hybrid micro-nano structured thermal interfaces for pool boiling heat transfer enhancement. Microelectron J 37:1158–1164

Maune H, Chiu HY, Bockrath M (2006) Thermal resistance of the nanoscale constrictions between carbon nanotubes and solid substrates. Appl Phys Lett 89:013109

Mudawar I, Anderson TM (1990) Parametric investigation into the effects of pressure, subcooling, surface augmentation and choice of coolant on pool boiling in the design of cooling systems for high-power-density electronic chips. J Electron Packag 112:375–383

Murad S, Puri IK (2008) Thermal transport across nanoscale solid-fluid interfaces. Appl Phys Lett 92:133105

Namburu P, Kulkarni D, Dandekar A, Das D (2007) Experimental investigation of viscosity and specific heat of silicon dioxide nanofluids. Micro Nano Lett 2:67–71 IET

Nelson IC, Banerjee D, Ponnappan R (2009) Flow loop experiments using polyalphaolefin nanofluids. J Thermophys Heat Transf 23:752–761

Sathyamurthi V, Ahn HS, Banerjee D, Lau SC (2009) Subcooled pool boiling experiments on horizontal heaters coated with carbon nanotubes. J Heat Trans 131:071501

Shanks HR, Maycock PD, Sidles PH, Danielson GC (1963) Thermal conductivity of silicon from 300 to 1,400 °K. Phys Rev 130:1743–1748

Shin D (2011) Molten salt nanomaterials for thermal energy storage and concentrated solar power applications. Dissertation, Texas A&M University, College Station

Shin D, Banerjee D (2011a) Enhancement of heat capacity of molten salt eutectics using inorganic nanoparticles for solar thermal energy applications. Dev Strateg Mater Comput Des II: Ceram Eng Sci Proc 32:119–126

Shin D, Banerjee D (2011b) Enhancement of specific heat capacity of high-temperature silica-nanofluids synthesized in alkali chloride salt eutectics for solar thermal-energy storage applications. Int J Heat Mass Transf 54:1064–1070

Shin D, Banerjee D (2011c) Enhanced specific heat of silica nanofluid. J Heat Trans 133:024501

Shin D, Banerjee D (2010) Effects of silica nanoparticles on enhancing the specific heat capacity of carbonate salt eutectic (work in progress). Int J Struct Changes Solids 2:25–31

Shin D, Banerjee D (2013) Enhanced specific heat capacity of nanomaterials synthesized by dispersing silica nanoparticles in eutectic mixtures. J Heat Transf (accepted, in print)

Singh N (2010) Computational analysis of thermo-fluidic characteristics of a carbon nano-fin. Dissertation, Texas A&M University, College Station

Singh N, Sathyamurthy V, Peterson W, Arendt J, Banerjee D (2010) Flow boiling enhancement on a horizontal heater using carbon nanotube coatings. Int J Heat Fluid Flow 31:201–207

Singh N, Unnikrishnan V, Banerjee D, Reddy J (2011) Analysis of thermal interfacial resistance between nanofins and various coolants. Int J Comput Methods Eng Sci Mech 12:254–260

Singh N, Shin D, Banerjee D (2012) In: Kaul AB (ed.) Nano-scale effects in multi-phase flows and heat transfer: microelectronics to nanoelectronics: materials, devices and manufacturability. CRC press (Taylor and Francis), Boca Raton, FL

Sriraman SR (2007) Pool boiling on nano-finned surfaces. Texas A&M University, College Station

Sunder M, Banerjee D (2009) Experimental investigation of micro-scale temperature transients in sub-cooled flow boiling on a horizontal heater. Int J Heat Fluid Flow 30:140–149

Ujereh S, Fisher T, Mudawar I (2007) Effects of carbon nanotube arrays on nucleate pool boiling. Int J Heat Mass Transf 50:4023–4038

Unnikrishnan VU, Banerjee D, Reddy JN (2008) Atomistic-mesoscale interfacial resistance based thermal analysis of carbon nanotube systems. Int J Therm Sci 47:1602–1609

Vajjha RS, Das DK (2009) Specific heat measurement of three nanofluids and development of new correlations. J Heat Transf 131

Wu W, Bostanci H, Chow L, Hong Y, Su M, Kizito J (2010) Nucleate boiling heat transfer enhancement for water and fc-72 on titanium oxide and silicon oxide surfaces. Int J Heat Mass Transf 53:1773–1777

Yu J (2012) Experimental investigation of heat transfer of nanofluids in a microchannel using temperature nanosensors. Dissertation, Texas A&M University, College Station

Zhou S-Q, Ni R (2008) Measurement of the specific heat capacity of water-based Al_2O_3 nanofluid. Appl Phys Lett 92: 093123–093123–093123

Zhou L-P, Wang B-X, Peng X-F, Du X-Z, Yang Y-P (2009) On the specific heat capacity of cuo nanofluid. Adv Mech Eng 2010

Chapter 2
Nanofins: Science

Abstract This chapter provides background information on carbon nanotubes (CNT), molecular dynamics (MD) simulations, and interfacial thermal resistance (R_k). The first section discusses carbon nanotubes, their structure, properties, and potential applications. The second section introduces the molecular dynamics simulations. This section discusses the basic equations describing the motion of the atoms, the force field to calculate potential energy of the system, boundary conditions, and ensembles employed in this work and space–time correlation to calculate the properties of the system. The last two sections detail the interfacial thermal resistance, theories developed to calculate interfacial resistance, importance of interfacial resistance at nanoscales, and the simulation techniques developed and used in calculating the interfacial thermal resistance.

Carbon Nanotubes

There are three known allotropes of carbon viz diamond, graphite, and fullerenes (or carbon nanotubes). The bonding hybridization of diamond is sp^3 while that of graphite, fullerenes, and carbon nanotubes is sp^2. Carbon nanotubes are formed by wrapping the two-dimensional planar graphite sheets. Graphite sheets can be single or multi-layer forming single and multi-wall carbon nanotubes, respectively. Also depending upon how a graphite sheet is rolled, a variety of nanotube structures can be formed. The physical structure of a carbon nanotube is uniquely defined by a vector (also known as "chirality") connecting two crystallographically equivalent sites on the 2D graphene sheet (Dresselhaus et al. 1992) known as chiral vector. Graphene is an atom thick sheet of sp^2 carbon atoms packed in a honeycomb crystal lattice. The chiral vector is expressed as

$$\vec{C}_h = n\hat{a}_1 + m\hat{a}_2 \tag{4}$$

N. Singh and D. Banerjee, *Nanofins*,
SpringerBriefs in Thermal Engineering and Applied Science,
DOI: 10.1007/978-1-4614-8532-2_2, © The Author(s) 2014

Figure 2.1 shows the unit vectors $\hat{a}_1$ and $\hat{a}_2$ and the (n, m) integer pair on a graphene sheet for defining different kinds of carbon nanotubes. Since the graphene lattice is rotationally symmetrical, n and m are considered such that $0 \le m \le n$. Each pair of (n, m) represents a unique structure of the nanotube, divided into three categories armchair, zigzag, and chiral. $m = 0$ corresponds to zigzag tubes, and $m = n$ corresponds to armchair tubes. All other tubes are known as chiral. The angle between the zigzag tube and the chiral vector is known as chiral angle θ. For zigzag tubes, the chiral angle is 0°, while for armchair, the chiral angle is 30°. For chiral nanotubes, $0 < \theta < 30°$ and is defined as

$$\theta = \sin^{-1} \frac{\sqrt{3}m}{2\sqrt{n^2 + nm + m^2}} \tag{5}$$

If the length of the carbon–carbon bond is known, the diameter of the nanotube is given by

$$d = \sqrt{3}a\sqrt{n^2 + nm + m^2} \Big/ \pi = C_h / \pi \tag{6}$$

where a is the carbon–carbon bond length and Ch is the length of the chiral vector $\overrightarrow{C_h}$

Figure 2.2 shows the formation of carbon nanotubes with different crystal structures from the graphene sheet.

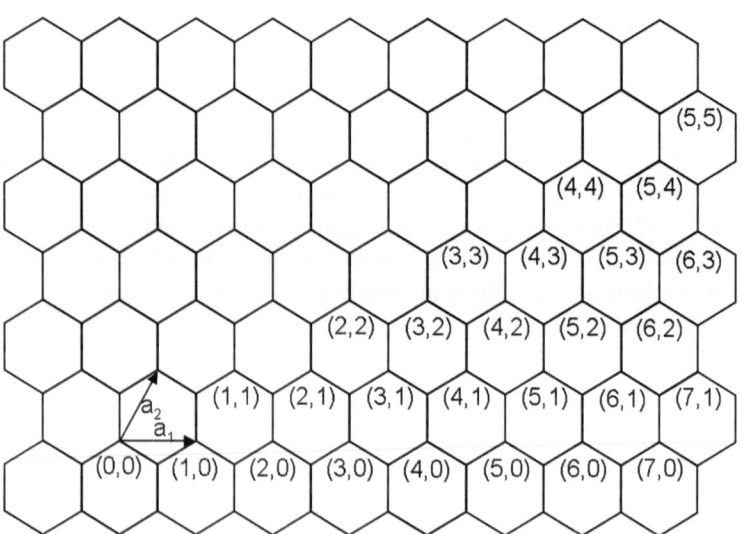

Fig. 2.1 Graphene sheet with chiral vectors and (n, m) indices

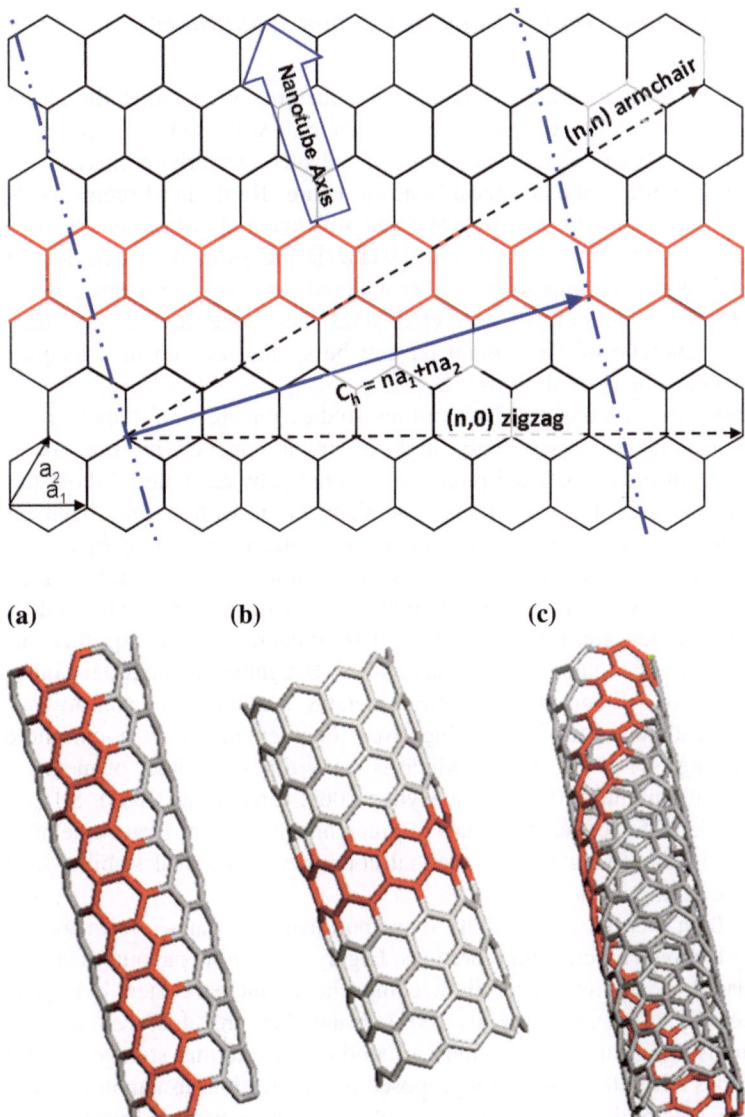

(a)   **(b)**   **(c)**

Fig. 2.2 Formation of carbon nanotubes from a 2D graphene sheet (*Top*) chiral vectors on the graphene sheet (*bottom*) carbon nanotube structures showing alignment of hexagons in **a** (5, 5) nanotube **b** (10, 0) nanotube and **c** (8, 1) nanotube

Properties and Applications of Carbon Nanotubes

The ability to manipulate the molecular structure by changing the chiral angle makes carbon nanotubes remarkable material with tunable properties. The extended network of C–C bonds results in enhanced material properties. The C–C bond of graphite is the strongest bond in nature (Ruoff and Lorents 1995). This makes carbon nanotubes very strong and stiff material with a tensile strength of 0.15TPa and a Young's modulus of 0.9TPa (Demczyk et al. 2002). So nanotubes find application in composites as filler materials to enhance properties like tensile strength and stiffness (Thostenson et al. 2005). Nanotubes have a very large aspect ratio. A nanotube of 1 nm diameter can be synthesized up to a length of few centimeters (Zheng et al. 2004).

Nanotubes are considered in various studies for electronic applications. The carbon nanotubes possess very high electrical conductivity due to the free movement of the delocalized pi-electron (donated by each atom) about the entire structure in a CNT. Graphite is a semiconductor with a zero band gap, but nanotubes can be either metallic or semiconductor depending upon the chiral vector. For n = m, armchair nanotubes are metallic in nature, whereas if n−m is a multiple of 3, the nanotube is semiconducting with a very small band gap. All other nanotubes are semiconductors with moderate band gap that inversely depends upon the diameter of the nanotubes (Baughman et al. 2002; Louie 2000). Due to the one-dimensional structure, metallic nanotubes are able to carry high currents without excessive heating over long lengths, with a current density approaching $\sim 4 \times 10^9$ A/cm^2, which is many times the order of metals such as copper and aluminum (Hong and Myung 2007; Liang et al. 2001). So nanotubes are considered to be ideal materials for interconnects in very large scale integration (VLSI) circuits owing to high thermal conductivity, thermal stability, and large current-carrying capacity.

The high thermal conductivity of carbon nanotubes arises from the high-frequency carbon–carbon bond vibrations. Highly conducting nanotubes are therefore considered ideal filler material in composite to increase their overall thermal conductivity (Thostenson et al. 2005). Nanotubes are also used to synthesize nanofluids which find applications in cooling and thermal storage technologies. Nanotubes have, therefore, been proposed for applications as nanofins to cool high heat flux electronic devices and as highly conducting thermal interface materials between microprocessor chips and heat sinks (Kordas et al. 2007; Xu et al. 2006, 2007).

Molecular Simulations

"Molecular simulation" is a generic term for theoretical methods and computational techniques based on advanced statistical tools to model or mimic the behavior of molecules. Molecular simulations provide exact solutions for the problems involving models based on statistical mechanics. The results of the molecular simulations depend solely on the nature of the theoretical method. In molecular simulations, the atomic system evolves from spatial and/or temporal domain in accordance with the extensive calculations of intermolecular energies. This is also the fundamental difference between the molecular simulations and the other computational methods.

There are two main families of molecular simulation methods: Monte Carlo (MC) and molecular dynamics (MD). Hybrid molecular methods are also developed, which combines features of these two techniques. Monte Carlo method is a stochastic technique that relies on the probabilities and random numbers. In a Monte Carlo simulation, a starting configuration domain is defined, then a new trail configuration is generated randomly by displacing, exchanging, removing, or adding a molecule and is evaluated against an acceptance criterion, which is based on the calculated change in energy or other property of the system, and at last, the trial configuration is either accepted or rejected based on the criterion. Since not all states contribute to configurational properties, states making most significant contributions are sampled by generating Markov chains. In Markov chain, the newly accepted state is always more favorable than the existing state. In contrast, molecular dynamics is a deterministic method in which time evolution of the system is determined by solving the equations of motion for each atom and/or molecule.

Both methods have their advantages and disadvantages depending upon the system being solved and the properties being simulated. MC simulations are time independent; they only define the state of the system at some point in time, whereas MD simulations define the time evolution of the system. Since the MD simulations solve equations of motion for each atom at each time step, these are computationally very costly, whereas in MC simulations, the atom movement is stochastically sampled, which is computationally cheap. But increase in the sample rejection rate can dramatically increase the computational cost in MC. As a result, MC method is better suited for studying systems of liquids with moderate densities, gases, systems with large number of coupled degree of freedom such as fluids, disordered materials, strongly coupled solids, and cellular structures. MD simulations are better suited for higher-density fluids, gases under high pressure and temperatures, and systems with dynamic properties such as transport coefficients, time-dependent responses to perturbations, rheological properties, and spectra (Allen 2004).

Molecular Dynamics Simulations

Molecular dynamics is a powerful computational technique to study the dynamical evolution of classical many body systems of solids, liquids, and gases. Classical laws of mechanics are obeyed by the motion of the atoms/molecules. This is a reasonably excellent approximation for wide range of systems and properties. MD simulations numerically integrate the Newton's equations of motion for all the atoms/molecules in the model system and output their temporal evolution of coordinates and momenta. Newton's equation of motion is a simple consequence of the Lagrange's equations of motion of classical mechanics. The temporal evolution is discrete, and at each time step, it is equivalent to a microstate in the same ensemble. Classical statistical mechanics equations can be used to derive the desired properties from these microstates. So MD simulations can be used to test hypothesis, characterize theories and experiments, simulate conditions impossible to achieve in experiments, and estimate missing or unreliable data. The power of MD lies in its ability to investigate atomic/molecular movements not accessible to experiments. Recent advances in algorithms and computers have made MD simulations feasible for large systems for longer timescales.

In general, Newton's equation of motion for an atom i can be expressed as

$$m\ddot{r}_i(t) = f_i(t) \tag{7}$$

Consider a system of N atoms, whose intermolecular potential energy is defined by $U(r^N)$, r^N representing the center of mass of the atoms $r^N = r_1, r_2, r_3 \ldots r_N$. The intermolecular potential energy of the system is defined by a set of parameters and their functional form, known as the force field. This is discussed in the next section. If no dissipative forces are acting among the atoms, for conserved intermolecular forces, the force acting on an atom i relates to the potential as

$$f_i(t) = -\frac{\partial U(r^N)}{\partial r_i} \tag{8}$$

Combining above two equations, we get

$$M\ddot{r}_i(t) = f_i(t) = -\frac{\partial U(r^N)}{\partial r_i} \tag{9}$$

Integrating the above equation once gives the atomic momentum at the next time step, and integrating twice yields the atomic positions. If the integration continues for a long time, we get the atomic momentum and position trajectories, which can then be used to obtain the macroscopic properties using the classical mechanics equations. The equations of motion are too complex to integrate for atomic systems, so numerical integration techniques are employed. There are number of schemes available, but the two most commonly used are Verlet algorithm and Gear predictor–corrector algorithm. In this work, Verlet algorithm is used and will be discussed. The Gear predictor–corrector scheme is discussed in details in the literature (Gear 1966, 1970, 1971a, b).

The Verlet algorithm is the most widely used method to integrate equations of motion initially adopted by Verlet (1967, 1968) and attributed to Stormer (Gear 1971b). This method gives direct solution of the Eq. (9). The basic idea is to calculate the positions at the next time step $(t + \delta t)$ from the previous $(t - \delta t)$ and the current time step (t). Writing the position $r(t)$ as Taylor series expansion from the previous and next time step,

$$r(t - \delta t) = r(t) + \dot{r}(t)\,\delta t - \frac{1}{2!}\ddot{r}(t)\,\delta t^2 + \frac{1}{3!}\dddot{r}(t)\,\delta t^3 + o(\delta t^4) \qquad (10)$$

$$r(t + \delta t) = r(t) + \dot{r}(t)\,\delta t + \frac{1}{2!}\ddot{r}(t)\,\delta t^2 + \frac{1}{3!}\dddot{r}(t)\,\delta t^3 + o(\delta t^4) \qquad (11)$$

Adding the above two equations lead to

$$r(t + \delta t) = 2 * r(t) - r(t - \delta t) + \ddot{r}(t)\,\delta t^2 + o(\delta t^4) \qquad (12)$$

In the above equation, $\ddot{r}(t)$ is calculated from Eq. (9). The truncation error is of the order of δt^4. The only problem is that there is no way to directly compute the velocities $(\dot{r}(t))$, which are used to calculate kinetic energy of the system and hence other quantities such as total energy, temperature, and pressure. The velocities are computed from the positions as

$$\dot{r}(t) = \frac{r(t + \delta t) - r(t - \delta t)}{2\,\delta t} + o(\delta t^2) \qquad (13)$$

However, the velocities are one step behind the positions and also the error associated is of the order of δt^2. The velocities at time $(t + \delta t)$ can be calculated as

$$\dot{r}(t) = \frac{r(t + \delta t) - r(t)}{\delta t} + o(\delta t) \qquad (14)$$

But now the accuracy is of the order of δt. To increase the accuracy of the velocities, this Verlet algorithm has been modified. One of the modified schemes is called leaf-frog (Hockney 1970). The leaf-frog method was not able to handle the velocities satisfactorily. So a velocity Verlet algorithm was proposed by Swope et al. (1982). This algorithm stores positions, velocities, and accelerations at the same time step and minimizes the round-off errors. The standard velocity Verlet algorithm is implemented as

$$r(t + \delta t) = r(t) + \dot{r}(t)\,\delta t + \frac{1}{2}\ddot{r}(t)\,\delta t^2 \qquad (15)$$

$$r(t + \frac{\delta t}{2}) = \dot{r}(t)\,\delta t + \frac{1}{2}\ddot{r}(t)\,\delta t \qquad (16)$$

$$\ddot{r}_i(t + \delta t) = -\frac{1}{m}\frac{\partial U(r)}{\partial r} \qquad (17)$$

$$r(t + \delta t) = \dot{r}(t + \frac{\delta t}{2}) + \frac{1}{2}\ddot{r}(t + \delta t)\,\delta t \qquad (18)$$

Using this algorithm, the positions, velocities, and accelerations are all computed at the same time step, but still order of the global error is δt^2. So choice of the time step is very crucial to the successful implementation of algorithm. A large time step will accrue errors as the simulation proceeds and a too small time step will increase the simulation time and round-off errors.

Force Field

Force field or potential energy functions are simplified mathematical equations to model interaction between the particles in a system. According to Brenner (2000), an effective force field should possess following four critical properties:

- Flexibility: A potential function must be flexible enough to accommodate a wide range of fitting data derived from experiments and ab-initio calculations.
- Accuracy: A potential function should be able to accurately reproduce an appropriate fitting database.
- Transferability: A potential function should be able to describe at least qualitatively, if not with quantitative accuracy, structures not included in a fitting database.
- Computational efficiency: The function evaluation should be computationally efficient depending upon the system size and timescales of interest as well as the available computer resources.

There are a large number of force fields available in the literature, which can be broadly grouped into generic force fields, biological force field, and class II force fields (Fried 2006). For this work, class II force field, polymer-consistent force field (PCFF), is used. Class II force fields make extensive use of anharmonic and cross-coupling terms to accurately represent the potential energy surface obtained from *ab initio* calculations. PCFF is derived from the CFF (Maple et al. 1988) developed by Biosym. Biosym merged with molecular simulations into the current company Accelrys (Accelrys).

The total potential energy of a system is the sum of the bonded and nonbonded interactions. The bonded interaction can be further represented as sum of valence and cross-terms.

$$
\begin{aligned}
E \text{ (total)} &= E \text{ (bond)} + E \text{ (nonbond)} \\
&= E \text{ (valence)} + E \text{ (cross-term)} + E \text{ (nonbond)}
\end{aligned}
$$

Figure 2.3 illustrates the various bonded interactions. The bonded valence energy consists of bond stretching, two bond angle bending, twisting (dihedral), and molecules going out of plane (oop).

$$
E \text{ (valence)} = E \text{ (bond)} + E \text{ (angle)} + E \text{ (torsion)} + E \text{ (oop)}
$$

The cross-terms interactions account for changes caused by the surrounding atoms, which include bond–bond, interaction between adjacent bonds; angle–angle, interaction between angles having common vertex atom; bond–angle, interaction between an angle and one of its bond; end bond torsion, interaction between a

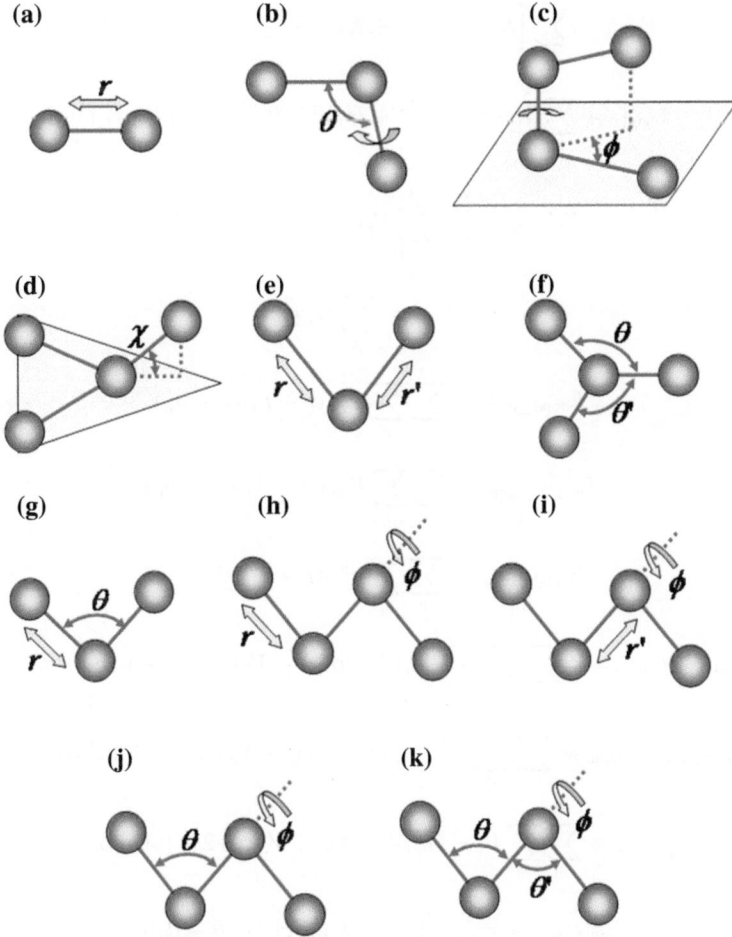

Fig. 2.3 Various types of bonded interactions between the atoms, considered in the PCFF force field

dihedral and one of its end bond; middle bond torsion, interaction between a dihedral and its middle bond; angle–torsion, interaction between a dihedral and one of its angles; angle–angle–torsion, interaction between a dihedral and its two valence angles.

$$E \text{ (cross-term)} = E \text{ (stretch--stretch)} + E \text{ (angle--angle)} + E \text{ (bond--angle)}$$
$$+ E \text{ (end\_bond-torsion)} + E \text{ (middle\_bond-torsion)}$$
$$+ E \text{ (angle-torsion)} + E \text{ (angle-angle-torsion)}$$

The equations representing these interactions in PCFF to calculate potential energy are defined as (Sun 1995),

$$E_{\text{bond}} = \sum_r \left[K_2(r - r_0)^2 + K_3(r - r_0)^3 + K_4(r - r_0)^4 \right] \tag{19}$$

$$E_{\text{angle}} = \sum_\theta \left[H_2(\theta - \theta_0)^2 + H_3(\theta - \theta_0)^3 + H_4(\theta - \theta_0)^4 \right] \tag{20}$$

$$E_{\text{torsiom}} = \sum_\phi \sum_{n=1}^{3} V_n(1 - \cos(n\phi - \phi_n^0)) \tag{21}$$

$$E_{\text{oop}} = \sum_\chi K_\chi \chi^2 \tag{22}$$

$$E_{\text{bond\_bond}} = \sum_r \sum_{r'} F_{rr'}(r - r_0)(r' - r_0') \tag{23}$$

$$E_{\text{angle\_angle}} = \sum_\theta \sum_{\theta'} F_{\theta\theta'}(\theta - \theta_0)(\theta' - \theta_0') \tag{24}$$

$$E_{\text{bond\_angle}} = \sum_r \sum_{\theta'} F_{r\theta}(r - r_0)(\theta - \theta_0) \tag{25}$$

$$E_{\text{end\_bond\_torsion}} = \sum_r \sum_\varphi (r - r_0)[V_1 \cos\phi + V_2 \cos 2\phi + V_3 \cos 3\phi] \tag{26}$$

$$E_{\text{middle\_bond\_torsion}} = \sum_r \sum_\phi (r' - r_0')[V_1 \cos\phi + V_2 \cos 2\phi + V_3 \cos 3\phi] \tag{27}$$

$$E_{\text{angle\_torsion}} = \sum_\theta \sum_\varphi (\theta - \theta_0)[V_1 \cos\phi + V_2 \cos 2\phi + V_3 \cos 3\phi] \tag{28}$$

$$E_{\text{angle\_angle\_torsion}} = \sum_\phi \sum_\theta \sum_{\theta'} K_{\phi\theta\theta'} \cos\phi(\theta - \theta_0)(\theta' - \theta_0') \tag{29}$$

The nonbonded interactions are composed of two types: van der Waals interactions and Coulomb's interactions.

$$E \text{ (nonbonded)} = E \text{ (LJ)} + E \text{ (coulomb)}$$

$$E_{LJ} = \sum_{i>j} E_{ij} \left[2\left(\frac{r_{ij}^0}{r_{ij}}\right)^9 - 3\left(\frac{r_{ij}^0}{r_{ij}}\right)^6 \right] \tag{30}$$

$$E_{\text{coulomb}} = \sum_{i>j} \frac{q_i q_j}{r_{ij}} \tag{31}$$

where q_i and q_j are partial charges, r_{ij}^0 is collision diameter (units of length), the distance at which the interparticle potential is zero, E_{ij} is the dissociation energy, and r_{ij} is the distance between the atoms. In PCFF, r_{ij}^0 and E_{ij} are calculated using the sixth power rule, as follows:

$$r_{ij}^0 = \left(\frac{\left(r_i^0\right)^6 + \left(r_j^0\right)^6}{2} \right)^{1/6} \tag{32}$$

$$E_{ij} = 2\sqrt{E_i E_j} \left(\frac{\left(r_i^0\right)^3 \times \left(r_j^0\right)^3}{\left(r_i^0\right)^6 + \left(r_j^0\right)^6} \right) \tag{33}$$

In a system with N particles, the numbers of pairwise interactions are N^2, which are computationally very costly. To minimize the computational cost, a cutoff radius, r_c, is defined. The atoms or molecules within the cutoff distance are only considered for the pairwise interactions as shown in Fig. 2.4.

Mathematically, this can be defined as

$$E_{\text{LJ,short-range}} = \begin{cases} V_{\text{LJ}}(r) - V_{\text{LJ}}(r_c) & r_{ij} \leq r_c \\ 0 & r_{ij} > r_c \end{cases} \tag{34}$$

Fig. 2.4 Nonbonded Lennard–Jones potential forces between two atoms and the use of cutoff radius to reduce the computational cost by eliminating calculations with atoms at a large distance

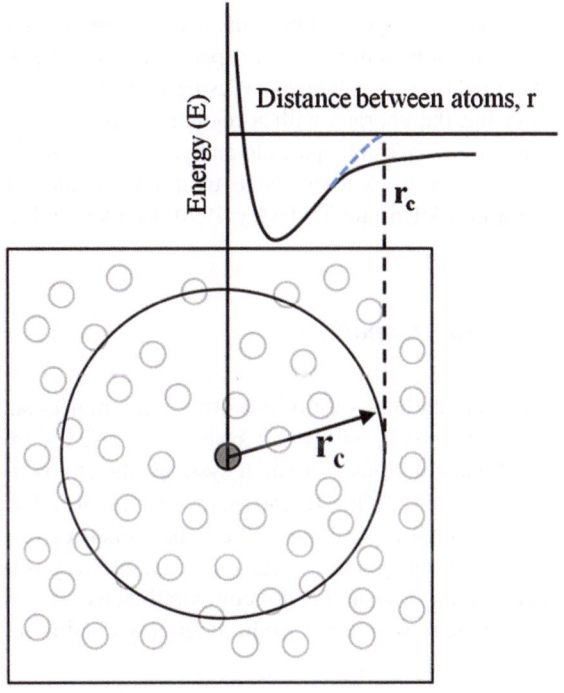

This technique is very efficient, but it completely ignores the interactions beyond the cutoff range. The LJ potential decays as r^{-6}, where r is the distance between the atoms. So it is acceptable to use cutoff distance to calculate potential energy and then add correction factor to account for the long-range interactions. The long-range interactions are given by (Allen and Tildesley 1990)

$$E_{LR} = 2\pi N \rho \int_{r_c}^{\infty} r^2 E_{LJ} dr \tag{35}$$

However, in case of Coulomb's interactions, the potential decays very slowly as r^{-1} leading to divergence of correction integrals. So for long-range coulombic interactions, other techniques are employed. One of the most popular methods is Ewald sum technique (Arnold and Holm 2005). For a neutral system of N particles with charges q_i and at position r_i, the total columbic energy is given as

$$E_{coulomb} = \frac{1}{2} \sum_{m}^{\infty} \sum_{i,j=1}^{N}{}' \frac{q_i q_j}{|r_{ij} + nL|} \tag{36}$$

where r_{ij} is the distance between i and j atoms, n is the count of periodic images, and L is the length of the unit cell assuming it is cubic in shape. Prime (') denotes that sum for $i = j$ will be ignored for the original box, $n = 0$. The energy given by the above equation strongly varies at small distances but slowly decays at large distances. A computational trick is to split the energy equation into two exponentially converging sums. This trick is computationally expensive as one part of the sum is solved in reciprocal space, thus needing several Fourier transformations. This method is accelerated by using fast Fourier transformation (FFT) method by replacing the charges with a regular mesh. The most common mesh-based technique used is particle–particle and particle–mesh (PPPM) method (Eastwood et al. 1977). A very comprehensive treatment of these methods can be found in the literature (Allen and Tildesley 1990; Frenkel and Smit 2002; Rapaport 2004).

Boundary Conditions

The computational cost of performing the molecular dynamics simulations is very high. As a result, only small systems with few atoms are typically considered for the MD simulations. For small systems, the atoms near the edges of the simulation box have few atoms to interact with. These edge atoms introduce the surface effects, which in turn affect the bulk properties of the system owing to its small size. To eliminate these surface effects, one way is to use a very large system. But solving a large system is not computationally feasible. The other way to overcome this problem is to use periodic boundary conditions.

With periodic boundary conditions, the simulation box is replicated infinitely in the space lattice. During the simulation, if one particle leaves the simulation box, another particle will enter the original box from the opposite face from the periodic image. This mechanism is demonstrated in Fig. 2.5 in a two-dimensional version. In 2D, the original box is surrounded by eight images of the box. As a particle crosses the right edge of the box, its image from the left edge enters the original box. So the original simulation box is devoid of any boundaries and thus eliminates the surface effects due to the presence of the boundaries. Apart from eliminating surface defects, periodic boundary conditions are useful in building the neighboring lists. As discussed above, neighboring lists contain the list of particles within the cutoff distance for each particle to calculate the nonbonded interactions. For the particles near the center of the simulation box, neighbors will be in the same box. But for a particle near the boundary, there are no particles to interact with beyond the boundary if periodic conditions are not used. With periodic boundary conditions, an image of the box is utilized, so the particle can interact with its neighbors similar to a particle at the center of the cell.

Ensembles

Ensemble is an idealization consisting of very large number of states of a system, considered all at once, one of which represents the state of the real system at any given instant. All possible states appear with an equal probability. So an ensemble is a system with different microscopic states representing an identical macroscopic or thermodynamics state.

Fig. 2.5 Methodology of implementation of periodic boundary conditions in order to reduce end effects

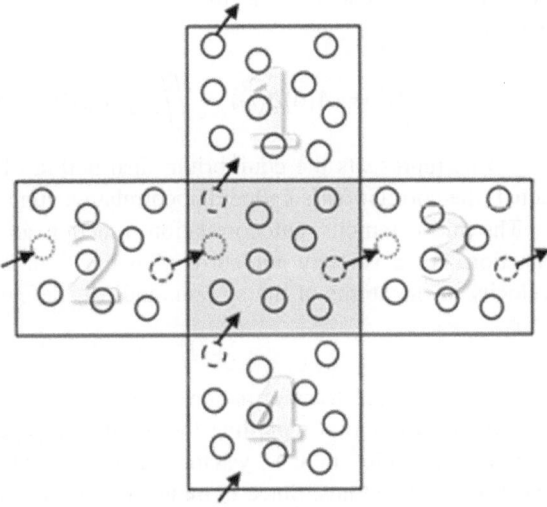

Newton's equations of motion explore only constant energy surfaces, characteristic of isolated systems. However, real systems interact with the surrounding, are exposed to pressure and external forces, and exchange thermal energy. These macroscopic constraints lead to different types of ensembles. The following ensembles are employed in this work.

- Microcanonical Ensemble (NVE): Collection of all the systems having constant total energy. This corresponds to a thermally isolated system. The thermodynamic state of the system is characterized by constant number of atoms, N, constant volume, V, and constant energy, E.
- Canonical Ensemble (NVT): Ensemble of systems, which exchange energy with a large heat reservoir. Heat capacity of the heat reservoir is large enough to maintain constant temperature for the coupled system. The thermodynamic state of the system is characterized by constant number of atoms, N, constant volume, V, and constant temperature, T.
- Isobaric-Isothermal Ensemble (NPT): Ensemble of systems, which exchange energy with a large heat reservoir of constant temperature, and the volume of systems changes in response to the applied external pressure. The thermodynamic state of the system is characterized by constant number of atoms, N, constant pressure, P, and constant temperature, T.

Space–Time Correlation Functions

Molecular dynamics simulations give the position, velocity, and acceleration of the atoms of a given system over time. So in essence, MD simulations give the apace and time evolution of the system. This time evolution of the system can then be used to determine a number of dynamic properties using the space–time correlation functions. If A and B are two variables dependent upon the momentum and position (p, q) of atoms, the time correlation function of variables A and B is defined as

$$C(t) = \langle A(0)B(t) \rangle = \iint f(p,q)A(p,q:0)B(p,q:t)\,dp\,dq \qquad (37)$$

$f(p,q)$ represents the equilibrium distribution of p and q. When A and B are equal, the correlation is called autocorrelation function.

The most famous autocorrelation function is the velocity autocorrelation function. For a velocity autocorrelation function, both A and B are equal to the velocity of the atoms of the system, and it is defined as

$$C(t) = \langle V(0).V(t) \rangle \qquad (38)$$

Taking Fourier transformation of the velocity correlation function reveals important information regarding the frequency-dependent properties such as diffusion of particles in the system, density of states of phonons, and vibration spectrum of the atoms. Since in molecular dynamics velocities are known at each time step, the complications related to the dependence of velocities on the position and momentum are avoided. So for the sake of calculations, the velocity auto-correlation function can be written as

$$C(t_k) = \langle V(0).V(t_k) \rangle \tag{39}$$

k subscript is added as the time is quantized into discrete steps. Also, usually in calculating the correlation function, a group of atoms is considered, and since the equilibration is done after minimization, the autocorrelation function does not need to start from time zero. Also, to better visualize the results, usually the autocor-relation function is normalized with respect to the initial value. Taking all these convections and approximations into account, for a group of N atoms, the nor-malized velocity autocorrelation function is defined as

$$\hat{C}(t_k) = \frac{C(t_k)}{C(0)} = \frac{\sum_{i=1}^{N} V_i(\tau).V_i(\tau + t)}{\sum_{i=1}^{N} V_i(\tau).V_i(\tau)} \tag{40}$$

Fourier transform of this velocity autocorrelation function represents the spectral density of the atomic motions, which can then used to determine the dominant modes of vibrations.

Interfacial Thermal Resistance

The interfacial thermal resistance represents a barrier to the heat flow at the boundary between two phases or two dissimilar materials. The existence of ther-mal resistance at the interface was first reported by Kapitza, with his measure-ments of the temperature drop at the interface between helium and a solid (Kapitza 1941). So the interface thermal resistance is also known as Kapitza resistance.

For a temperature drop of ΔT at the interface, the heat flux, J_Q, is related to the Kapitza resistance, R_k as

$$R_k = \frac{\Delta T}{J_Q} \tag{41}$$

The inverse of interface resistance R_k is called interfacial conductance, which is analogous to bulk conductivity in a macroscopic system. For a matrix system with constant thermal conductivity λ, the Kapitza length L_k is defined as

$$L_k = R_k \lambda \tag{42}$$

Kapitza length is the equivalent thickness of matrix, with the same thermal conductivity λ, over which the drop in temperature is same as the drop at the interface. Kapitza length gives an idea of the relative importance of the interfacial resistance.

Khalatnikov (1952) presented a model known as acoustic mismatch (AM) model, to explain and estimate the thermal resistance at the boundaries to helium.

According to this model, interfacial thermal resistance is due to the mismatch in the acoustic impedances of two dissimilar materials at the interface. Mazo and Onsager (1955) proposed the modern form of AMM. They applied acoustic theory and approximated liquid helium and the solid as a continuous elastic medium. Classical wave propagation equations were used to determine the transmission and reflection coefficients for phonons. Plane wave solutions founded in the two mediums were matched at the boundary. By imposing displacement and stress boundary conditions and assuming no scattering takes place at the interface, transmission coefficient, t_{AB}, for phonon energy in material A incident normal to the interface of material B is given by

$$t_{AB} = \frac{4Z_A Z_B}{(Z_A + Z_B)^2} \tag{43}$$

where $Z = \rho c$ is acoustic impedance, ρ is the mass density, and c is the speed of sound in the material (Mazo 1955). The transmission coefficient is the ratio of the total flux in the transmitted wave to the incident wave. Using this model, the authors predicted the boundary resistance to be of the order of 1,000 cm^2 K/W at 1 K and varies as T^{-3}. Little (1959) extended the acoustic mismatch model for to solid–solid interface. They showed that for a perfectly joined interface, the heat flow is proportional to the difference between the fourth powers of the temperature on each side of the interface, but results deviate for rough surfaces and for surfaces pressed into contact with one another.

Experimentalists observed that the Kapitza resistance is two orders small than predicted by AMM theory at 1 K, but thermal resistance between solids was higher than calculated for the AMM theory proposed by Little. Lee and Fairbank (1959) and Anderson et al. (1964) measured the boundary resistance between cooper and He^3 and found similar magnitude and temperature dependence to that found for a copper–superfluid He^4 boundary. Anderson et al. also found lower magnitudes of boundary resistance between solid He^3 or liquid He^4 and copper. According to AMM boundary, resistance is a strong function of pressure; however, the authors found that the boundary resistance is affected by pressure only near 0.1 K; around 1 K, it is independent of pressure. Challis et al. (1961) proposed transmission increase due to the existence of compressed layer of liquid on the surface of the solid formed due to the van der Waals attraction of liquid atoms to the solid. Matsumoto et al. (1977) measured thermal resistance across cooper–epoxy–cooper sandwiches over a temperature range of 0.05–10 K. Their data also supported the suggestion that low-frequency phonons contribute to thermal transport across the sandwich. They showed that thermal resistance between solid–solid interfaces can match AMM theory results depending upon the surface preparation. Shiren et al. (1981) found that resistance of a defective interface is much higher than a clean one. Swartz et al. (2009) measured the boundary resistance from 0.6 to 200 K between metal films and dielectric substrates onto which the films were deposited. The authors demonstrated that at low temperatures, below 40 K, the results were in good agreement with the AM model, but at higher temperature, the results vary significantly. They also found that other than carefully prepared surfaces, phonons in the frequency range above a few gigahertzes strongly scattered at all the surfaces. So in order to estimate the effect of diffuse scattering, they proposed the diffuse mismatch model (DMM). In contrast to AM, in diffuse mismatch (DM) model, all phonons striking an interface scatter to one side or other side of interface with the probability that is proportional to the phonon density of states. Thus, both models assume the interface without any intrinsic properties and so the structure of the interface does not affect the energy transmitted. In essence, these theories assume phonon transmission to be completely specular without scattering or diffuse scattering without any dependence on angle of incidence. So neither of the two models is able to predict the precise value of the interfacial resistance, but they provide a good reference against which to compare the experimental results.

Simulations of Interfacial Thermal Resistance

Experimentally interfacial thermal resistance between dissimilar materials has been studied extensively since its first introduction by Kapitza. Based on the experimental results, AMM and DMM theories were developed to explain the phenomenon of interface thermal resistance. Experiments results have shown that the interface resistance is affected by the surface conditions, but these theories do not take surface conditions into account. Numerical simulations have been used to fill in the gap between the experimental results and the theoretical analysis. Maris and others (Pettersson and Mahan 1990; Young and Maris 1989) used traditional lattice dynamic simulations to study the thermal boundary resistance between two dissimilar solids. Numerical calculations were performed to obtain the phonon transmission coefficient and group velocity, which were then used to obtain the phonon flux density across the interface. From the phonon flux density, Kapitza resistance is obtained. But these calculations are limited only to simple interfaces and are not a viable approach for grain boundary problems. To handle complex problems like gain structures, molecular dynamics simulations were devised. Molecular dynamics methods were well established for the calculations of thermal conductivity. Though both equilibrium, Green–Kubo (Gillan and Dixon 1983; Green 1954; Kubo 1966), and nonequilibrium, direct methods (Jund and Jullien 1999; Muller-Plathe 1997) in molecular dynamics technique can be used to calculate the thermal conductivity of the materials at the atomistic level, only nonequilibrium methods can be employed to calculate the interfacial thermal resistance.

To calculate the interfacial thermal resistance using nonequilibrium molecular dynamics (NEMD) simulations, a thermal current is created in the system. There are two approaches to create a thermal current in the system. The first approach is the constant heat flux method in which equal amount of energy is added and removed from two plates of the system as shown in Fig. 2.6.

The most popular methods used to add or remove energy from the system were developed by Jund and Jullien (1999). In this method, the energy is added or

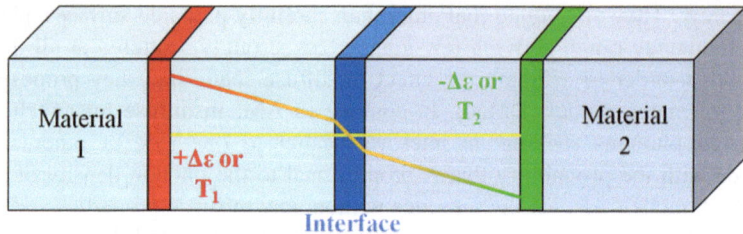

Fig. 2.6 Schematic used for performing the nonequilibrium molecular dynamic simulations. $\Delta\varepsilon$ denotes the heat added or removed from the system in constant heat flux simulations and T_1, and T_2 denotes the fixed high and low temperatures for constant temperature simulations. *Yellow line (horizontal)* denotes the initial (base) temperature of the system

removed from the system by scaling the instantaneous velocities of the atoms at discrete time steps as

$$\vec{v}_i = \vec{v}_g + \alpha\left(\vec{v}_i - \vec{v}_g\right) \tag{44}$$

where $\vec{v}_g$ is the velocity of the center of mass of the particles in the plate and

$$\alpha = \sqrt{1 \pm \frac{\Delta\varepsilon}{KE}} \tag{45}$$

+ sign is used for atoms in the hot plate and − sign for atoms in the cold plate, $\Delta\varepsilon$ is the amount of energy added or subtracted from the plates and usually taken as 1 % of $k_B T$ to introduce only small temperature gradients within the system. The relative kinetic energy is given by

$$KE = \frac{1}{2}\sum_{i=1}^{n} m_i \vec{v}_i^2 - \frac{1}{2}\sum_{i=1}^{n} m_i \vec{v}_g^2 \tag{46}$$

A second method to add and remove energy from the system was proposed by Muller-Plathe (1997). In this method, the velocity of the fastest atom in the cold plate is exchanged with the velocity of the slowest atom in the hot plate. In this way, energy is added to the hot plate and same amount of energy is removed from the cold plate. Additional calculations to sort the velocities have to be performed to get the slowest and the fastest atoms.

The second approach to create thermal current in the system is by keeping the two plates at different temperatures. One plate is kept at higher constant temperature and the other at lower constant temperature. The difference in the temperatures leads to a thermal current in the system as heat flows from hot plate to cold plate. As the heat flows from hot plate to cold plate, the temperature of hot plate decreases and that of cold plate increases. The temperature of the plates is kept constant either by scaling the velocities of the atoms in the plate or by using a temperature thermostat. Murad et al. (2008) used Gaussian thermostat to keep the temperatures of the plates constant while simulate the effect of pressure and surface wetting on the interfacial thermal resistance between the Si crystal and water molecules. Cummings et al. (2004) used the velocity scaling method to keep the temperatures of the plates constant while studying the thermal conductivity in Y-junction nanotubes and the effect of defects in straight nanotubes.

In velocity scaling procedure, the desired temperature is reached by scaling instantaneous velocities of the atoms in a plate at discrete time steps as

$$v_{i,\text{new}} = v_{i,\text{old}} \sqrt{\frac{T_{\text{desired}}}{T_{\text{current}}}} \tag{47}$$

where $v_{i,\text{new}}$ is the new velocity of the ith atom, $v_{i,\text{old}}$ is the old velocity of the ith atom, T_{current} and T_{desired} are current and desired temperatures of the plate carrying the ith atom. The amount of energy added to the hot plate and subtracted from the cold plate is equivalent to the net change in the kinetic energy of the atoms in that plate and is given as

$$\Delta E_{\text{plate}} = \frac{1}{2} \sum_{i=1}^{n} m_i (\vec{v}_{i,\text{new}}^2 - \vec{v}_{i,\text{old}}^2) \tag{48}$$

where m_i is the mass of ith atom in the plate, n is the number of atoms in the plate, assuming all the atoms in a given plate have same mass, i.e., for monoatomic systems. Then, the heat flux in the system is calculated from the average energy added or removed from the hot and cold plates over the period of the data collection as

$$J = \frac{1}{2A} \frac{\sum_{i=1}^{N} |\Delta E_{\text{hotplate}} + \Delta E_{\text{coldplate}}|}{t} \tag{49}$$

where N is the number of velocity rescaling operations performed during the simulations, A is the cross-sectional area of the system, perpendicular to the flow of heat flux, and t is the total time of the simulation. Since the hot and cold plates are situated in the middle of the material 1 and material 2, as shown in Fig. 2.6, the heat flows in both the directions. So a factor of 1/2 is added to the above equation, assuming heat equally divides in both the directions. In both these approaches, once the equilibrium is reached, the thermal conductivity of a material can be obtained using the Fourier law. Also, from the temperature drop at the interface and the heat flux in the system, interfacial thermal resistance can be obtained using Eq. (2.41).

Originally, these methods were developed to calculate the bulk thermal conductivity of the materials, but have been modified to calculate the interfacial thermal resistance. Maiti et al. (1997) performed the first nonequilibrium MD simulations to study the interfacial thermal resistance. This method was based on the approach used by Tenenbaum et al. (1982) to calculate the thermal conductivity of the materials. They studied Kapitza resistance on the Si grain boundaries by creating thermal current within the system using constant temperature plates. They also demonstrated that local equilibrium can be achieved in a small system by running simulations for a very long time. Schelling et al. (2004) proposed an approach similar to that used by Maiti et al. (1997) to calculate the Kapitza resistance between three twisted Si grain boundaries. They found that increase in the structural disorder at the grain boundary leads to a significant increase in the Kapitza resistance. Xue et al. (2003) studied the liquid–solid interaction using the constant temperature model. They found out that magnitude of the intermolecular forces between the liquid and the solid atoms plays a key role in determining the interfacial thermal resistance.

Xiang et al. (2008) studied the role of the interface on interface resistance in microscale and nanoscale heat transfer processes using MD simulations. They found that the thermal contact resistance increases exponentially with decreasing area of the microcontact and also increases with increasing microcontact layer thickness. They also found that the material defects increase the thermal resistance. Wang et al. (2008) studied cross-plane thermal conductivity and interfacial thermal resistance between bilayered films using nonequilibrium MD simulations. The bilayer films consist of argon atoms and another material identical to argon but with different atomic mass. The results showed large temperature jump at the interface suggesting important role of interfacial resistance to heat transfer. They found interfacial resistance to be dependent on the mass ratio of the atoms but independent of the film thickness. Murad et al. (2008) studied the effect of pressure and liquid properties on the interfacial thermal resistance. They found that high fluid pressure and hydrophilic surfaces facilitate better acoustic matching and thus decrease the interface resistance.

Maruyama et al. (2004) performed the first molecular dynamic simulations to study the interfacial thermal resistance between carbon nanotubes in a bundle. They calculated the resistance as 6.46×10^{-8} m^2 K/W. Zhong and Lukes (2006) studied the dependence of the interfacial thermal resistance between two carbon nanotubes as a function of nanotube spacing, overlap, and length. From the simulations, they found that there was a fourth order of magnitude reduction in the interfacial thermal resistance if the nanotubes are bought into intimate contact. The resistance also reduced for longer nanotubes and nanotubes with more overlap area. They attributed the increase to the increase in area available for heat transfer between the nanotubes. Greaney et al. (2007) employed MD simulations to elucidate the factors responsible for heat transfer between carbon nanotubes. Their calculations showed that sharp resonance between nanotubes allows efficient energy transfer.

Huxtable et al. (2003) made transient absorption measurements on individual single-walled carbon nanotubes encased in cylindrical micelles of sodium dodecyl sulphate (SDS) surfactant, dispersed in D$_2$O to study nanotube–matrix interface resistance. Ti:sapphire mode-locked laser was used to produce a series of subpicosecond pulses with wavelength in the range of 740–840 nm. The decay constant of 45 ps was calculated for the heat flow from the carbon nanotube to SDS corresponding to interfacial resistance of 8.04×10^{-8} m^2 K/W. Shenogin et al. (Kapitza 1941) studied the interfacial thermal resistance between the carbon nanotubes and octane molecules surrounding it. They found a large value of interfacial thermal resistance, and the resistance was found to decrease as the length of the carbon nanotubes increased and then stabilized to a constant value. They attributed this large resistance to the weak coupling between the rigid tube structure and the soft organic liquid. Clancy et al. (2006) studied the effect of chemical functionalization of carbon nanotubes on the interfacial thermal resistance and thermal conductivity of nanocomposites. They grafted linear hydrocarbon chains on the nanotube surface using covalent bonds. They found decrease in the interfacial thermal resistance between functionalized nanotubes and the surrounding polymer matrix.

Unikrishnan et al. (2008) studied the effect of chemical additives like CuO on the carbon nanotubes suspended in water. There was a marginal increase in the resistance with the addition of impurities in the water. Recently, Carlborg et al. (2008) studied the thermal boundary resistance between single-walled carbon nanotube and matrices of solid and liquid argon by performing classical molecular dynamics simulations. They found that the resistance does not depend on the length of the nanotube longer than 20 Å. Also, they suggested that resonant coupling between the low-frequency modes of the carbon nanotube and the argon matrix are responsible for heat transfer from the carbon nanotube to the matrix molecules.

Numerical Methodology

These simulations are based on the lumped capacitance analysis. This analysis is applicable to any system which has a very small Biot number. Biot number is defined as

$$B_i = \frac{hL}{k} \tag{50}$$

where h is the heat transfer coefficient at the interface, L is the characteristic length of the system, and k is the thermal conductivity of the solid. In essence, this analysis is applicable to systems in which the thermal gradients within solid body are negligibly small as it cools with time, i.e., body has a very high thermal conductivity. Also, the characteristic length is small and the coefficient of heat transfer from the solid body to the surrounding fluid is also small.

Carbon nanotubes have very high thermal conductivity. So the temperature gradients within the carbon nanotube are negligibly small. Consider a system with carbon nanotube at temperature T_{hot} surrounded by fluid with temperature T_{cold}. The pool of liquid is large enough such that it remains at T_{cold}. Let m be the mass of the nanotube, c is its specific heat, A_s the surface area, and G the interfacial conductance (inverse of interfacial resistance), then mathematically,

$$mc\frac{dT}{dt} = -GA_s(T - T_{cold}) \tag{51}$$

At time $t = 0$, the temperature of the object is T_{hot} and cold fluid is T_{cold}, and after time t, the temperature of the hot object is T and of cold fluid as T_{cold}, integrating the above equation from time $t = 0$ to t, we get

$$(T - T_{\text{cold}}) = (T_{\text{hot}} - T_{\text{cold}}) \exp\left[-\frac{GA_s}{mc}t\right] \tag{52}$$

Or this equation can be written as

$$\Delta T(t) = \Delta T_{\text{initial}} \exp\left[-\frac{t}{\tau}\right] \tag{53}$$

where τ is the thermal time constant. The solution indicates that the difference in the temperature of solid and fluid, $\Delta T(t)$, approaches zero as the time approaches infinity. The quantity $C = mc/A_s$, heat capacity per unit area, is constant for carbon nanotubes (Huxtable et al. 2003). So, once we know the time constant, we can find the interfacial thermal resistance using the following equation:

$$R_k = \frac{\tau}{C} \tag{54}$$

Figure 2.7 shows the simulation procedure to calculate the interfacial thermal resistance using molecular dynamics simulations. Each step is described below. All the simulations are performed using the LAMMPS software (LAMMPS 2008; Plimpton 1995). LAMMPS stands for large-scale atomic/molecular massively parallel simulator. LAMMPS is a classical open source molecular dynamics simulation code to be used as particle simulator at the atomic, meso-, or continuum scale. It was originally developed in FORTRAN and later ported to C++. It uses message-passing interface (MPI) protocol and is designed to run efficiently on parallel computers.

Problem Setup

The first step in the simulation is to prepare the starting structure. The starting structure is prepared by placing the carbon nanotube at the centre of the simulation box. The axis of the nanotube is aligned along the z-axis. The matrix molecules surround the carbon nanotube. The density of the unit cell is same as the density of the bulk matrix at room temperature.

Figure 2.8 shows a typical starting structure. The starting structure is created using Materials Studio (MS), a commercially available software. The initial structure created by amorphous builder module of MS is not suitable for these simulations, as the nanotube axis is not aligned in the z-direction. So the structure is manually edited to align the nanotube along the z-axis and to move any matrix molecules inside the nanotube. This structure is exported, and a script file converts the MS files to LAMMPS input file. This file acts as the starting input structure for LAMMPS.

Fig. 2.7 Numerical
procedure to calculate the
interfacial thermal resistance
employed in this work

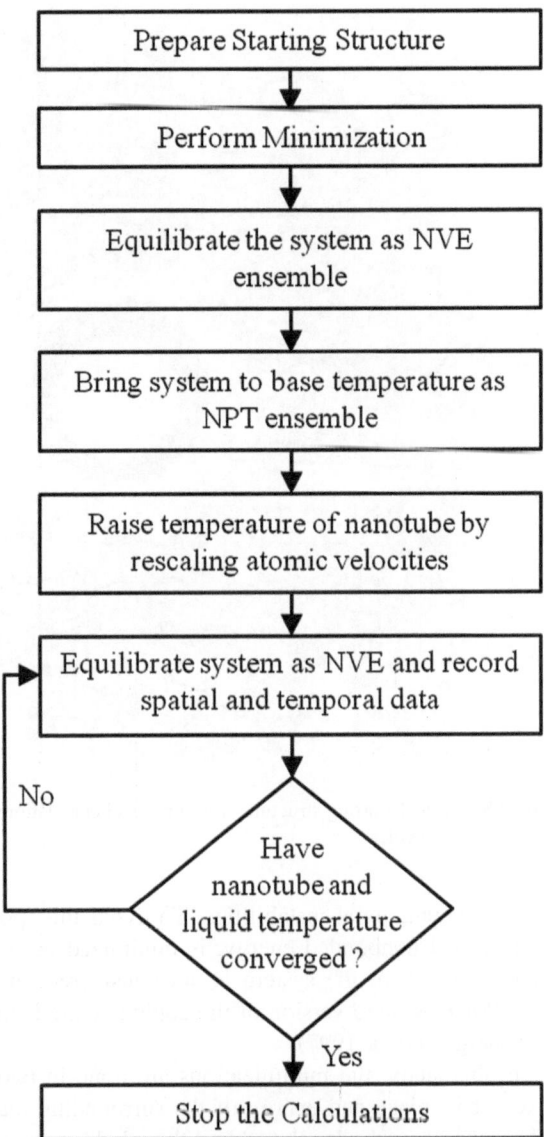

Minimization

The starting structure is formed by randomly placing the molecules inside the
simulation box. The atoms of the matrix molecules may be very close to or
overlapping with atoms of carbon nanotube or other matrix molecules. If simulations are performed with this starting structure, very large forces will be exerted
on the atoms, as a result imparting very high velocities to the atoms. This will

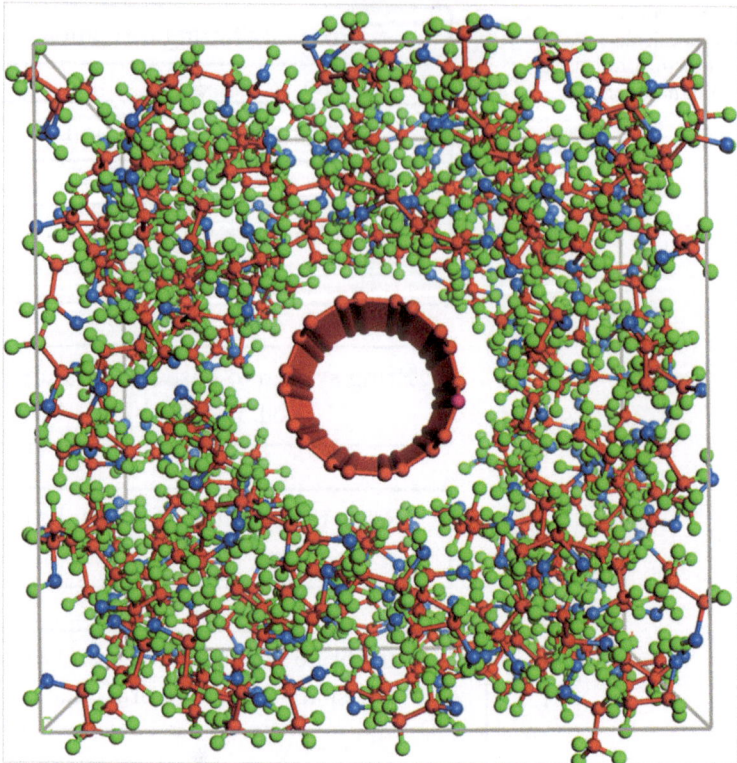

Fig. 2.8 Typical starting structures for numerical calculations with carbon nanotube surrounded by matrix molecules

cause computational instabilities. To avoid this, potential energy of the system, bonded and nonbonded energy, is minimized before running the dynamics. The kinetic energy of the system is not considered during minimization. LAMMPS uses Polak–Ribiere version of the conjugate gradient (CG) algorithm to minimize the energy (Polak 1997).

In this study, the minimizations are done in two steps. First, the nanotube is fixed at its place and energy of the surrounding matrix molecules is minimized. Then, the nanotube is relaxed and the whole system is allowed to minimize. First step is necessary, as during the structure preparation, some matrix molecules are too close to the nanotube. The proximity of these matrix molecules will dent the nanotube and distort it beyond the point of recovery. So in the first step, only the matrix molecules will move. Once the matrix molecules are at a distance from the nanotube, the nanotube is relaxed. In this way, in the final structure, the nanotube is not significantly distorted.

Equilibration

Minimization is followed by equilibration. After minimization, the system is theoretically at 0 K. To start dynamics, the system has to be brought to base temperature of interest, 300 K. For this work, this is done in two steps. In the first step, velocities are assigned to the atoms randomly at 300 K, and the system is allowed to relax as an NVE ensemble for 50 ps. As the initial configuration (coordinates) is not an equilibrium one, during relaxation some of the kinetic energy is converted to potential energy. So the final temperature of the system is lower than 300 K. In all the simulations, the final temperature converges to a value close to 160 K. In the second step, the temperature of the system is raised from 160 to 300 K by equilibrating the system as an NPT ensemble for 25 ps, followed by equilibrating at 300 K for 250 ps at a pressure of 1 atmosphere. The pressure and the temperature of the system are controlled using the Nose–Hover thermostat (Hoover 1985) and barostat (Hoover 1986). During equilibration, pressure of 1 atmosphere is applied on the sides of the simulation box that responds independently of the pressure component along the axis of the tube. Thus, the box length can change in x–direction and y-direction, but not in the z-direction.

Production Run

After the system is equilibrated to a uniform temperature of 300 K, the temperature of the nanotube is instantaneously increased to 700 K in case of n-heptane and its variants and 750 K in case of other matrix systems by simple velocity scaling procedure. If T_{base} is the base temperature and $T_{desired}$ is the desired temperature, the new velocities of the atoms, v_{new}, from the old velocities v_{old} are given by

$$v_{new} = v_{old} \sqrt{\frac{T_{desired}}{T_{base}}} \tag{55}$$

Subsequently, the system is allowed to relax under constant energy as an NVE ensemble. As the simulation proceeds, the energy from the nanotube is slowly transferred to the matrix molecules. Snapshots of the temperature distribution within the box are obtained at fixed intervals of 100 steps. Also, after every 100 steps, the average temperature of the nanotube and matrix molecules is calculated and stored. The simulation is terminated as the temperature of the carbon nanotube approaches the liquid temperature.

The numerical models involving noncontinuum assumption—discussed in this chapter—can be used to study the behavior of a nanofin to further elucidate the nanofin effect. This is discussed in the next chapter.

References

Allen MP (2004) Introduction to molecular dynamics simulation. Comput Soft: Synth Polym Proteins 23:1–28

Allen MP, Tildesley DJ (1990) Computer simulation of liquids. Clarendon Press, New York

Anderson AC, Connolly JI, Wheatley JC (1964) Thermal boundary resistance between solids and helium below 1 k. Phys Rev 135:A910–A921

Arnold A, Holm C (2005) Efficient methods to compute long-range interactions for soft matter systems. Adv Polym Sci 185:59–109

Baughman RH, Zakhidov AA, de Heer WA (2002) Carbon nanotubes-the route toward applications. Sci 297:787–792

Brenner DW (2000) The art and science of an analytic potential. Comput Simul Mater At Level 217:23–40

Carlborg CF, Shiomi J, Maruyama S (2008) Thermal boundary resistance between single-walled carbon nanotubes and surrounding matrices. Phys Rev B 78:205406

Challis LJ, Dransfeld K, Wilks J (1961) Heat transfer between solids and liquid helium *ii*. Proc R Soc Lond A 260:31–46

Clancy TC, Gates TS (2006) Modeling of interfacial modification effects on thermal conductivity of carbon nanotube composites. Polymer 47:5990–5996

Cummings A, Osman M, Srivastava D, Menon M (2004) Thermal conductivity of y-junction carbon nanotubes. Phys Rev B 70:115405

Demczyk BG, Wang YM, Cumings J, Hetman M, Han W, Zettl A, Ritchie RO (2002) Direct mechanical measurement of the tensile strength and elastic modulus of multiwalled carbon nanotubes. Mater Sci Eng, A 334:173–178

Dresselhaus MS, Dresselhaus G, Saito R (1992) Carbon fibers based on c60 and their symmetry. Phys Rev B 45:6234–6242

Eastwood J, Hockney R, Lawrence D (1977) P3m3dp: the three-dimensional periodic particle–particle/particle-mesh program. Comput Phys Commun 19:215–261

Frenkel D, Smit B (2002) Understanding molecular simulation: from algorithms to applications. Academic Press, Orlando

Fried JR (2006) Computational parameters. Physical Properties of Polymers Handbook, pp 59–68

Gear C (1966) The numerical integration of ordinary differential equations of various orders. ANL-7126, Argonne National Lab, Ill

Gear CW (1970) The simultaneous numerical solution of differential-algebraic equations. Stanford University, Stanford

Gear CW (1971a) The automatic integration of ordinary differential equations. Commun ACM 14:176–179

Gear CW (1971b) Numerical initial value problems in ordinary differential equations. Prentice Hall PTR, NJ

Gillan MJ, Dixon M (1983) The calculation of thermal conductivities by perturbed molecular dynamics simulation. J Phys C: Solid State Phys 16:869–878

Greaney PA, Grossman JC (2007) Nanomechanical energy transfer and resonance effects in single-walled carbon nanotubes. Appl Phys Lett 98:125503

Green MS (1954) Markoff random processes and the statistical mechanics of time-dependent phenomena. ii. Irreversible processes in fluids. J Chem Phys 22:398–413

Hockney RW (1970) The potential calculation and some applications. Methods Comput Phys 9:136–211

Hong S, Myung S (2007) Nanotube electronics: a flexible approach to mobility. Nat Nanotechnol 2:207–208

Hoover WG (1985) Canonical dynamics: equilibrium phase-space distributions. Phys Rev A 31:1695–1697

Hoover WG (1986) Constant-pressure equations of motion. Phys Rev A 34:2499–2500

Huxtable ST, Cahill DG, Shenogin S, Xue L, Ozisik R, Barone P, Usrey M, Strano MS, Siddons G, Shim M (2003) Interfacial heat flow in carbon nanotube suspensions. Nat Mater 2:731–734

Jund P, Jullien R (1999) Molecular-dynamics calculation of the thermal conductivity of vitreous silica. Phys Rev B 59:13707 13711

Kapitza PL (1941) The study of heat transfer in helium ii. J Phys (USSR) 4:181–210

Khalatnikov IM (1952) Discontinuities and large amplitude sound waves in helium ii. Zh Eksp Teor Fiz 23:253–260

Kordas K, Toth G, Moilanen P, Kumpumaki M, Vahakangas J, Uusimaki A, Vajtai R, Ajayan P (2007) Chip cooling with integrated carbon nanotube microfin architectures. Appl Phys Lett 90:123105

Kubo R (1966) The fluctuation-dissipation theorem. Rep Prog Phys 29:255–284

LAMMPS (2008). Available at http://lammps.Sandia.Gov/index.Html

Lee DM, Fairbank HA (1959) Heat transport in liquid he[3]. Phys Rev 116:1359–1364

Liang W, Bockrath M, Bozovic D, Hafner JH, Tinkham M, Park H (2001) Fabry-Perot interference in a nanotube electron waveguide. Nature 411:665–669

Little WA (1959) The transport of heat between dissimilar solids at low temperatures. Can J Phys 37:334–349

Louie SG (2000) Electronic properties, junctions, and defects of carbon nanotubes. Top Appl Phys 80:113–146

Maiti A, Mahan GD, Pantelides ST (1997) Dynamical simulations of nonequilibrium processes— heat flow and the kapitza resistance across grain boundaries. Solid State Commun 102:517–521

Maple JR, Dinur U, Hagler AT (1988) Derivation of force fields for molecular mechanics and dynamics from ab initio energy surfaces. Proc Natl Acad Sci 85:5350–5354

Maruyama S, Igarashi Y, Taniguchi Y, Shibuta Y (2004) Molecular dynamics simulations of heat transfer issues in carbon nanotubes, The 1st International Symposium on Micro and Nano Technology Honolulu, Hawaii

Matsumoto DS, Reynolds CL Jr, Anderson AC (1977) Thermal boundary resistance at metal-epoxy interfaces. Physical Review B 16:3303–3307

Mazo RM (1955) Theoretical studies on low temperature phenomena. Yale University, New Haven

Muller-Plathe F (1997) A simple nonequilibrium molecular dynamics method for calculating the thermal conductivity. J Chem Phys 106:6082–6085

Murad S, Puri IK (2008) Thermal transport across nanoscale solid-fluid interfaces. Appl Phys Lett 92:133105

Pettersson S, Mahan GD (1990) Theory of the thermal boundary resistance between dissimilar lattices. Phys Rev B 42:7386–7390

Plimpton S (1995) Fast parallel algorithms for short-range molecular dynamics. J Comput Phys 117:1–19

Polak E (1997) Optimization: algorithms and consistent approximations

Rapaport DC (2004) The art of molecular dynamics simulation

Ruoff RS, Lorents DC (1995) Mechanical and thermal properties of carbon nanotubes. Carbon 33:925–930

Schelling PK, Phillpot SR, Keblinski P (2004) Kapitza conductance and phonon scattering at grain boundaries by simulation. J Appl Phys 95:6082–6091

Shiren NS (1981) Surface roughness contribution to kapitza conductance. Phys Rev Lett 47:1466–1469

Sun H (1995) Ab initio calculations and force field development for computer simulation of polysilanes. Macromolecules 28:701–712

Swartz ET, Pohl RO (2009) Thermal resistance at interfaces. Appl Phys Lett 51:2200–2202

Swope WC, Andersen HC, Berens PH, Wilson KR (1982) A computer simulation method for the calculation of equilibrium constants for the formation of physical clusters of molecules: Application to small water clusters. J Chem Phys 76:637–649

Tenenbaum A, Ciccotti G, Gallico R (1982) Stationary nonequilibrium states by molecular dynamics. Fourier's law. Phys Rev A 25:2778–2787

Thostenson ET, Li C, Chou TW (2005) Nanocomposites in context. Compos Sci Technol 65:491–516

Unnikrishnan VU, Banerjee D, Reddy JN (2008) Atomistic-mesoscale interfacial resistance based thermal analysis of carbon nanotube systems. Int J Therm Sci 47:1602–1609

Verlet L (1967) Computer experiments on classical fluids. I. Thermodynamical properties of Lennard-Jones molecules. Phys Rev 159:98–103

Verlet L (1968) Computer experiments on classical fluids. ii. Equilibrium correlation functions. Phys Rev 165:201–214

Wang S, Liang X (2008) Thermal conductivity and interfacial thermal resistance in bilayered nanofilms by nonequilibrium molecular dynamics simulations. Int J Thermophys :1–10

Xiang H, Jiang PX, Liu QX (2008) Non-equilibrium molecular dynamics study of nanoscale thermal contact resistance. Mol Simul 34:679–687

Xu Y, Zhang Y, Suhir E, Wang X (2006) Thermal properties of carbon nanotube array used for integrated circuit cooling. J Appl Phys 100:074302

Xu Y, Leong CK, Chung DDL (2007) Carbon nanotube thermal pastes for improving thermal contacts. J Electron Mater 36:1181–1187

Xue L, Keblinski P, Phillpot SR, Choi SUS, Eastman JA (2003) Two regimes of thermal resistance at a liquid-solid interface. J Chem Phys 118:337–339

Young DA, Maris HJ (1989) Lattice-dynamical calculation of the kapitza resistance between fcc lattices. Phys Rev B 40:3685–3693

Zheng LX, O'Connell MJ, Doorn SK, Liao XZ, Zhao YH, Akhadov EA, Hoffbauer MA, Roop BJ, Jia QX, Dye RC et al (2004) Ultralong single-wall carbon nanotubes. Nat Mater 3:673–676

Zhong H, Lukes JR (2006) Interfacial thermal resistance between carbon nanotubes: molecular dynamics simulations and analytical thermal modeling. Phys Rev B 74:125403

Chapter 3
Nanofins: Applications

Abstract In this chapter, the results from the non-equilibrium molecular dynamics simulations are presented. The consequences of the nanofin effect are presented—such as surface adsorption of solvent molecules leading to density oscillations and effects on the resultant material properties are discussed.

Non-equilibrium molecular dynamic simulations were performed to study the effect of chemistry and molecular structure on the interfacial thermal resistance between a carbon nanotube and coolant molecules. The coolants considered in this study are water, ethyl alcohol, 1-hexene, *n*-heptane, and its dimers, trimers, and their isomers. To study the effect of the chemical composition of the coolants on the thermal interfacial resistance, the molecules considered are water, ethyl alcohol, 1-hexene, and *n*-heptane. Also, *n*-heptane, *n*-tridecane, and *n*-nonadecane are studied to study the effect of polymer chains (structural variations) on the interfacial resistance. Also, isomers of *n*-heptane, *n*-tridecane, and *n*-nonadecane are considered to study the effect of branching and mixtures of chains on the interfacial resistance.

Figure 3.1 shows the molecular structure of matrix systems under consideration.

Molecular Structure

Liquid near the solid surface is much more ordered and denser than the bulk liquid, and therefore, the properties are akin to that of solid than a liquid. The liquid density exhibits molecular level oscillation in the direction normal to the liquid–solid interface (Yu et al. 2000). With distance from the interface, the density peaks to a large value followed by small oscillations and then approaches to the bulk density. The magnitude of the layering extends to a few atomic distances and strongly depends upon the solid–liquid bonding strength. This trend of ordering and fluctuations is seen in these simulations. All the coolants considered in this

N. Singh and D. Banerjee, *Nanofins*,
SpringerBriefs in Thermal Engineering and Applied Science,
DOI: 10.1007/978-1-4614-8532-2_3, © The Author(s) 2014

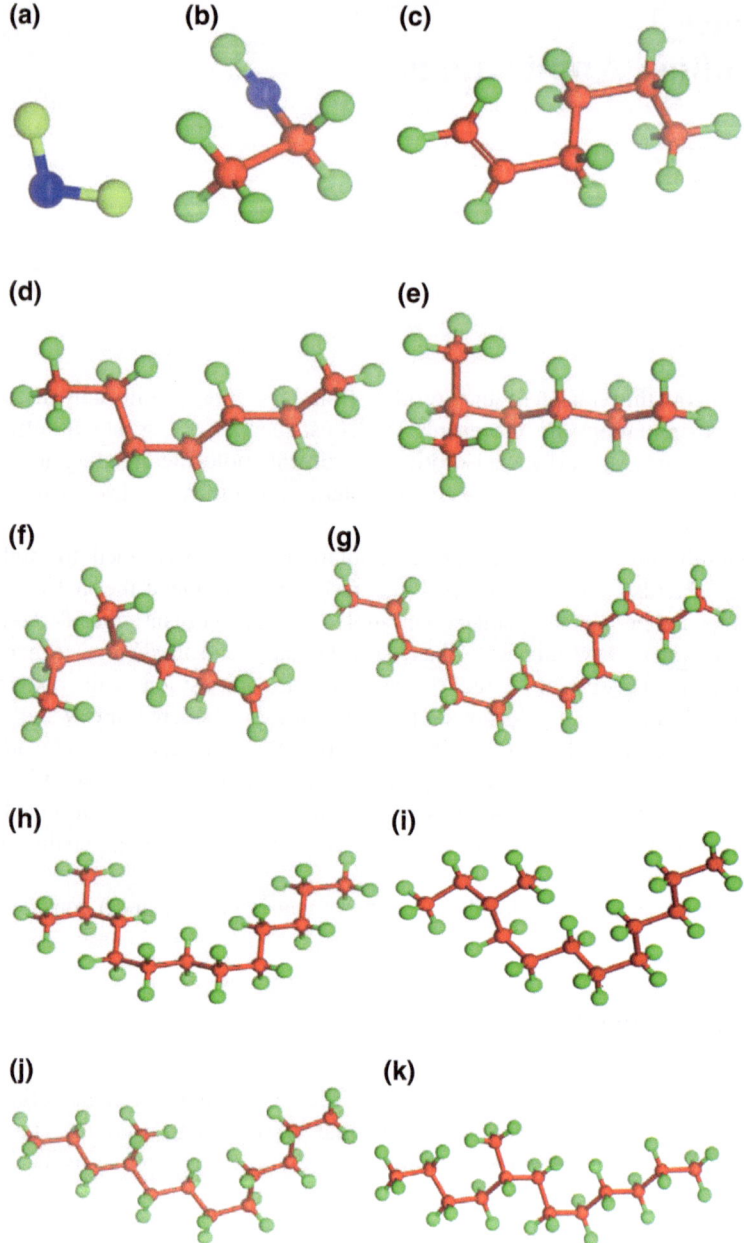

Fig. 3.1 Structure of molecules considered in this work. The atoms in *red color* are carbon, *green* hydrogen, and *blue* oxygen. **a** Water. **b** Ethyl alcohol. **c** 1-hexene. **d** *n*-heptane. **e** 2-methyl hexane. **f** 3-methyl hexane. **g** *n*-tridecane. **h** 2-methyl dodecane. **i** 3-methyl dodecane. **j** 4-methyl dodecane. **k** 5-methyl dodecane. **l** 6-methyl dodecane. **m** *n*-nonadecane. **n** 2-methyl octadecane. **o** 3-methyl octadecane. **p** 4-methyl octadecane. **q** 5-methyl octadecane. **r** 6-methyl octadecane. **s** 7-methyl octadecane. **t** 8-methyl octadecane. **u** 9-methyl octadecane

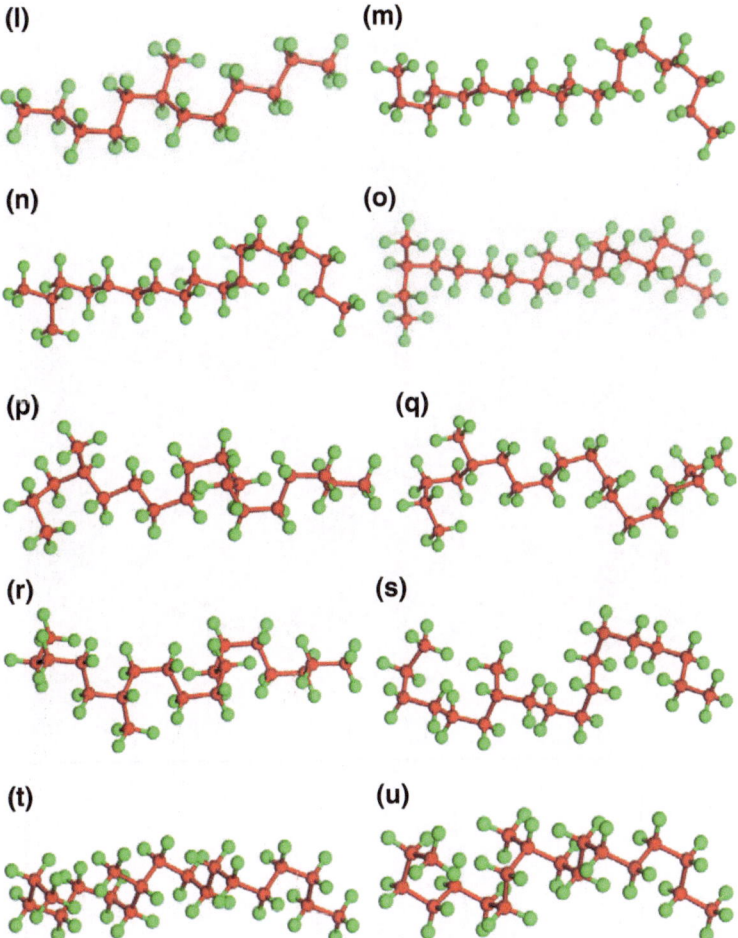

Fig. 3.1 continued

study exhibit the molecular ordering and density fluctuations near the nanotube wall. Figure 3.2 shows the molecular structure and the density fluctuations for *n*-tridecane molecule. In the molecular structure, we see clearly a gap of about 3 Å between the carbon nanotube and the surrounding coolant molecules due to the presence of the weak van der Waals forces. Also from the figure, it is clear that the atoms are densely packed at a distance of about 7 Å. This distance corresponds to the first peak density. From the spatial variation in the density, the figure shows that the oscillations and later the density settle down to bulk density of the liquid at the room temperature to about 0.75 g/cc^3.

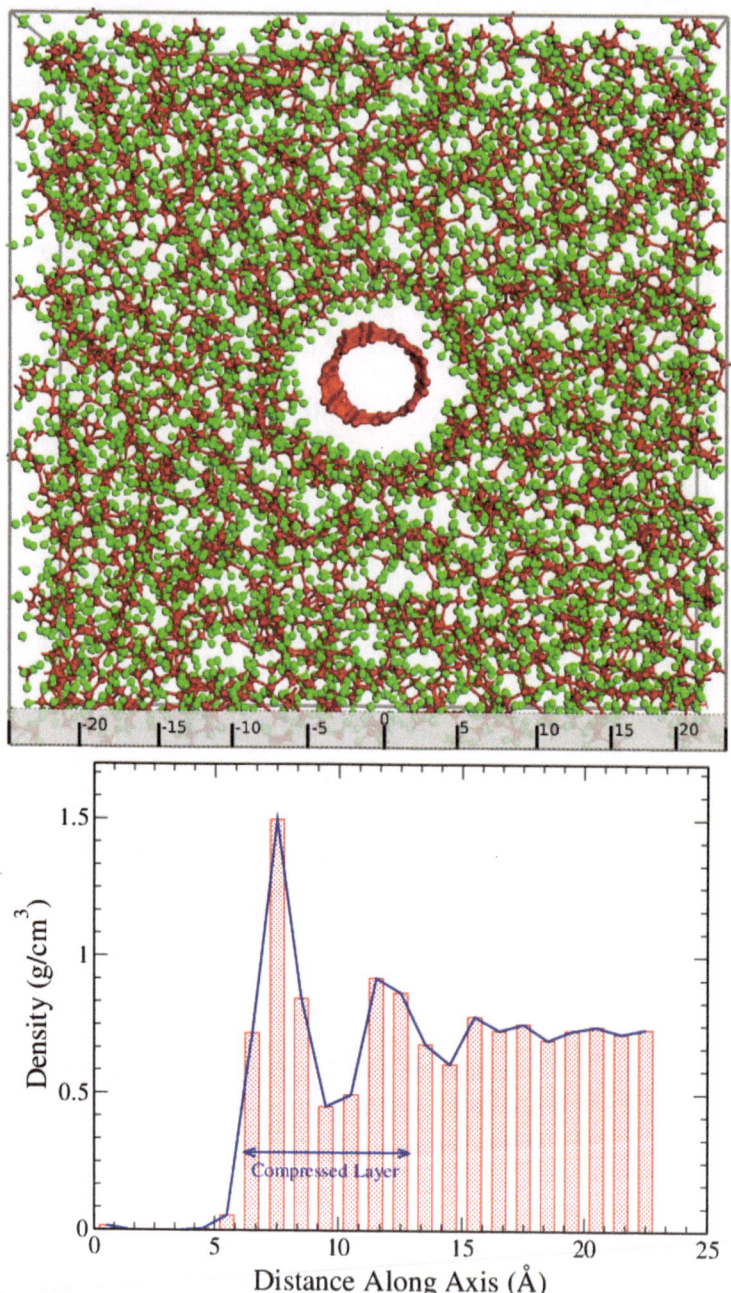

Fig. 3.2 *Top* Molecular structure of carbon nanotube surrounded by *n*-tridecane molecules at equilibrium. *Bottom* Radial variation of coolant density within the simulation box

Temporal Temperature Distribution

As discussed above, these simulations are based on the lumped capacitance method. The main prerequisites of the lumped capacitance with reference to these simulations are that there should not be any temperature gradients within the carbon nanotube. The very large thermal conductivity of the carbon nanotube addresses this requirement. The other prerequisites are that the temperature of the fluid surrounding the nanotube should remain constant and should not vary significantly with time. Also, the temporal decay of the temperature difference between the carbon nanotube and surrounding fluid is expected to be exponential. To ascertain these assumptions, temporal distribution of the temperature between the carbon nanotube and coolants is studied. Figure 3.3 shows the typical variation of the temperature in the carbon nanotube and surrounding fluid molecules. From the figure, it is clear that the temperature of the fluid molecules remains constant. Also the temperature difference between the nanotube and the fluid is observed to decay exponentially.

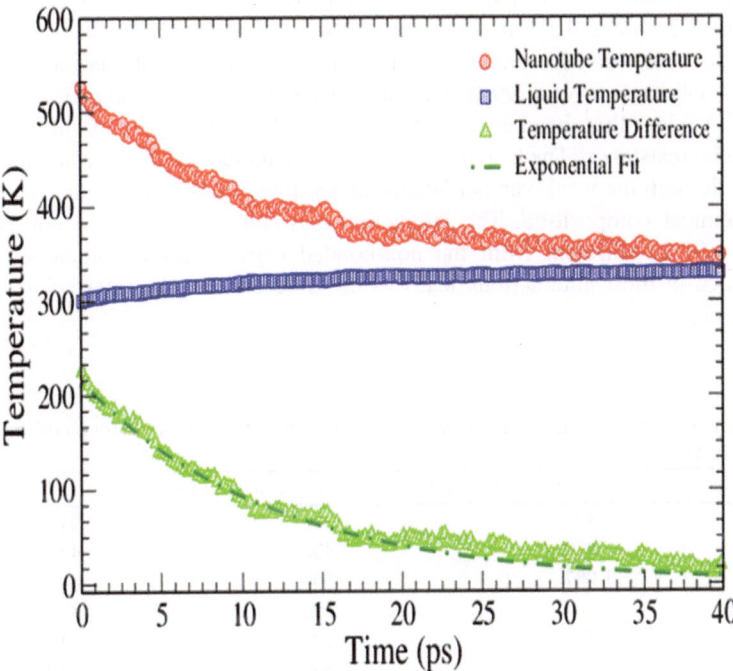

Fig. 3.3 Typical temporal variation of temperature of carbon nanotube and surrounding water molecules during a production run

Effect of Chemistry

To study the effect of chemistry on the interfacial thermal resistance water, ethyl alcohol and 1-hexene molecules are considered. Table 3.1 shows the parameters of the simulations that were used in this study. The cross-sectional area of the simulation box in all these simulation is 26.3×26.3 Å. The length of the nanotube is varied to study whether the length of the nanotube has any affect. It was observed that with small nanotubes, the value of the interfacial resistance fluctuates but becomes steady for large length.

Figure 3.4 shows the temporal variation of temperature difference of carbon nanotube and fluids in this study on a semi-logarithmic scale. Since the temporal decay of temperature difference is exponential, on the semi-logarithmic scale, the results are straight lines. Inverse of the slope of the lines give the time constant for the temperature decay. The interfacial thermal resistance at nanotube–liquid interface is calculated as

$$R_k = \frac{\tau A_{\text{cnt}}}{C_T} \tag{56}$$

where τ is the relaxation time constant, A_{cnt} is the area of the nanotube, and C_T is the heat capacity of the nanotube. The heat capacity of the nanotube per unit area is usually taken as a constant, 5.6×10^{-4} J/m^2 K (Huxtable et al. 2003).

Table 3.2 shows the time constant and interfacial thermal resistance for the molecules under consideration. The intervals of the interfacial thermal resistance have a confidence level of 99 % calculated using the regression analysis. From the table, it is clear that the chemical composition (chemistry) of the fluid affects the interfacial resistance. The heat transfer from the nanotube to the surrounding fluids occurs through the weak van der Waals interactions. These interactions depend on the chemical composition. The interaction between non-bonded carbon–oxygen atoms will be different from the non-bonded carbon–carbon interactions. The difference in these interactions leads to the difference in the interfacial thermal resistance.

Table 3.1 Parameters of molecules considered to study the effect of chemistry on the interfacial thermal resistance

Matrix system	No. of matrix molecules	No. of atoms in CNT	CNT length (Å)
Water	435	200	24.6
	870	400	49.19
	1,087	500	61.49
	1,523	700	86.04
Ethyl alcohol	123	200	24.6
	308	400	49.19
1-hexene	64	200	24.6
	133	400	49.19

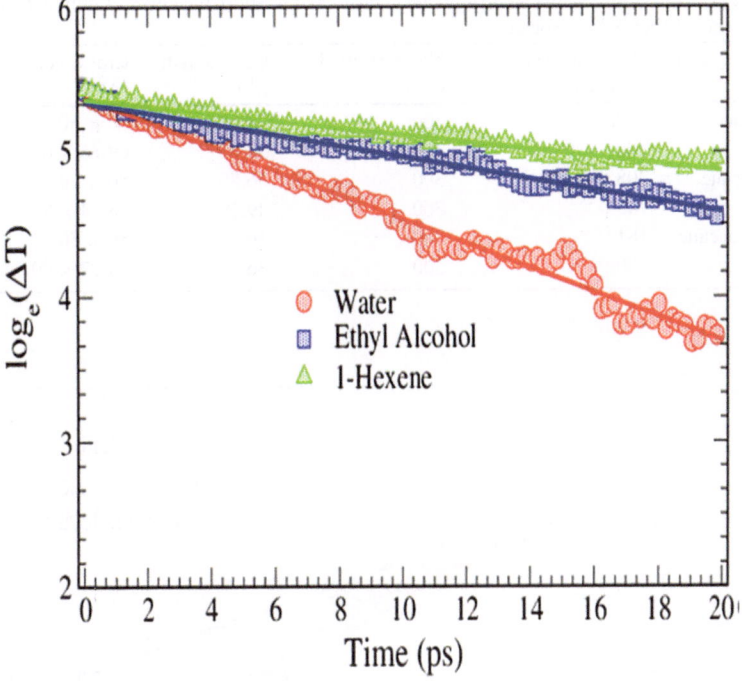

Fig. 3.4 Decay in temperature difference with time for water, ethyl alcohol, and 1-hexene

Table 3.2 The effect of chemistry on the relaxation time constant and interfacial thermal resistance

Matrix system	Relaxation time constant τ (ps)	Interfacial resistance $R_k \times 10^8$ (m^2K/W)
Water	11.9	$2.13^{+0.04}_{-0.04}$
Ethyl alcohol	26.5	$4.74^{+0.10}_{-0.11}$
1-hexene	40.8	$7.29^{+0.20}_{-0.20}$

Effect of Polymer Chains

Hydrocarbons formed by the catalytic oligomerization of poly-alpha-olefins (PAO) usually consist of long-chain hydrocarbons like n-heptane and their dimers, trimers, and tetramers. In this work, n-heptane, along with its dimer (n-tridecane) and trimer (n-nonadecane) has been studied to elucidate the effect of polymer chains on the interfacial thermal resistance.

Table 3.3 shows the parameters considered in this work.

Figure 3.5 shows the temporal variation of temperature difference between the carbon nanotube and fluids on a semi-logarithmic scale.

Table 3.3 Molecule systems and parameters considered to study the effect of polymer chains on the interfacial thermal resistance

Matrix system	No. of matrix molecules	No of atoms in CNT	CNT length (Å)	Cross-sectional area (Å²)
n-heptane	229	200	24.6	50 × 50
	700	300	36.89	69.86 × 69.86
n-tridecane	208	300	36.89	50 × 50
	749	360	49.19	88.74 × 88.74
n-nonadecane	181	300	36.89	50 × 50
	700	300	36.89	99.97 × 99.97

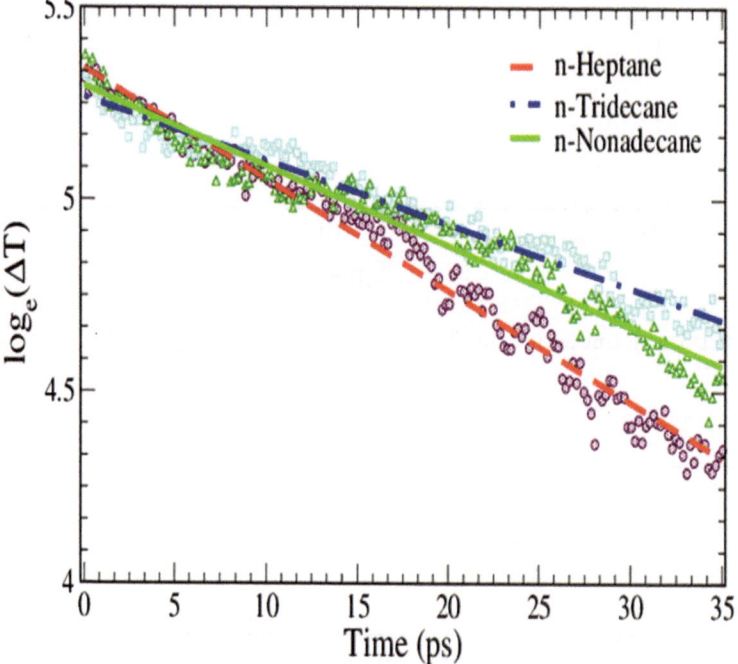

Fig. 3.5 Temporal decay of temperature on a semi-logarithmic scale for *n*-heptane, *n*-tridecane, and *n*-nonadecane

Table 3.4 shows the interfacial thermal resistance between the carbon nanotube and the fluids considered in this work. From the results, it is clear that polymer chains affect the interfacial thermal resistance. Since the polymers chains are formed from the same type of atoms, the mechanism of heat transfer (the weak van der Waal's forces between the atom types) will be same for all the molecules. Hence, it can be expected that chemical composition cannot be the sole cause for the observed variation in the interfacial resistance.

Table 3.4 Effect of polymer chains on the relaxation time constant and interfacial thermal resistance	Matrix system	Relaxation time constant τ (ps)	Interfacial resistance $R_k \times 10^8$ (m^{2}K/W)
	n-heptane	35.6	$6.35^{+0.11}_{-0.12}$
	n-tridecane	57.1	$10.20^{+0.19}_{-0.14}$
	n-nonadecane	45.9	$8.19^{+0.26}_{-0.22}$

From the results for n-heptane, the base molecule has the lowest resistance. This is due to the fact that n-heptane is able to efficiently wrap around the carbon nanotube due to small carbon chain. In case of n-tridecane and n-nonadecane, the chains are large and are not able to wrap around the nanotube as efficiently as n-heptane (stearic hindrance). Also n-nonadecane has a very long chain, which at equilibrium bends to form a chain with two legs at 90°. This type of configuration more efficiently wraps around the nanotube than a single straight chain like in the case of n-tridecane.

Effect of Isomers and Their Mixtures

To study the effect of isomers on the interfacial thermal resistance, isomers of n-heptane, n-tridecane, and n-nonadecane are considered in this work. Since in the isomers, the atoms of the compound are organized in different ways, the change in the structure can affect the interfacial resistance. Also the hydrocarbons produced by the catalytic oligomerization of PAO results in a mixture of hydrocarbons—consisting of their isomers, dimers, trimers, etc. The presence of mixtures may reduce the interfacial thermal resistance as the chains of different lengths and orientations may fit together much better (i.e., the stearic hindrance is reduced) than that of the liquid phase or solvent with only a single species of these compounds. The calculated time constant and hence the interfacial thermal resistances using Eq. (54) are shown in Table 3.5. In case of n-heptane and its isomers, n-heptane has lower interfacial thermal resistance than its isomers 2-methyl hexane and 3-methyl hexane. For n-tridecane molecule, the interfacial thermal resistance is higher than its isomers.

Figure 3.1 shows the equilibrium molecular structure of n-heptane, n-tridecane, and their isomers. From the equilibrium structure, for n-heptane and its isomers, the branching of the methane group leads to stearic hindrance and thus leading to ineffective wrapping of molecules around the carbon nanotube. In case of n-tridecane and its isomers, the resistance decreases as the methyl group advances from first carbon atoms to adjacent atoms. As seen from the equilibrium molecular structures, n-tridecane has a long straight chain, whereas its isomers form a branched structure. Due to the branching, the isomers effectively wrap around the carbon nanotubes. As a result, the resistance to energy transfer from carbon nanotubes to the molecules decreases significantly. Result of the interfacial

Matrix system	Relaxation time constant τ (ps)	Interfacial resistance $R_k \times 10^8$ (m^2K/W)
Table 3.5 Relaxation time constant and interfacial thermal resistance for different molecules and their isomers		
n-heptane	35.6	$6.35^{+0.11}_{-0.12}$
2-methyl hexane	43.2	$72.2^{+0.08}_{-0.08}$
3-methyl hexane	49.5	$8.85^{+0.12}_{-0.13}$
Mixture heptane & isomers	47.4	$8.46^{+0.09}_{-0.10}$
Tridecane	57.1	$10.20^{+0.19}_{-0.14}$
2-methyl dodecane	51.8	$9.25^{+0.113}_{-0.09}$
3-methyl dodecane	49.9	$8.91^{+0.11}_{-0.11}$
4-methyl dodecane	42.1	$7.52^{+0.12}_{-0.11}$
5-methyl dodecane	56.3	$10.06^{+0.15}_{-0.16}$
6-methyl dodecane	37.5	$6.70^{+0.11}_{-0.12}$
Mixture of tridecane and isomers	39.5	$7.06^{+0.1}_{-0.1}$
Nonadecane	45.9	$8.19^{+0.22}_{-0.26}$
2-methyl octadecane	48.5	$8.67^{+0.10}_{-0.13}$
3-methyl octadecane	55.9	$9.98^{+0.30}_{-0.28}$
4-methyl octadecane	46.1	$8.23^{+0.16}_{-0.17}$
5-methyl octadecane	38.5	$6.87^{+0.15}_{-0.12}$
6-methyl octadecane	44.2	$7.90^{+0.18}_{-0.08}$
7-methyl octadecane	48.8	$8.71^{+0.12}_{-0.12}$
8-methyl octadecane	36.0	$6.42^{+0.12}_{-0.14}$
9-methyl octadecane	50.8	$9.06^{+0.07}_{-0.16}$
Mixture of nonadecane and isomers	38.6	$6.89^{+0.16}_{-0.16}$
	41.7	$7.44^{+0.10}_{-0.10}$

thermal resistance for 5-methyl dodecane is anomalous, as the resistance is much higher than the expected results.

In case of *n*-nonadecane and its isomers, it is difficult to deduce a trend in the interfacial resistance. We can see oscillations, as resistance increases, then decreases, and again increases and decreases as the methyl group moves away from the first carbon atom. These oscillations are attributed to the relative effectiveness of these isomer chains in wrapping around the carbon nanotube. Consider the molecular structure, 5-methyl octadecane has a branched structure while 6-methyl octadecane is nearly a straight chain. So 5-methyl octadecane would wrap around the carbon nanotube effectively and should have lower interfacial resistance than 6-methyl octadecane. As expected, this trend can be seen from the simulation results in Table 3.5.

The mixtures also show deviation in the interfacial thermal resistance from the single-molecule systems. In all the cases, however, the resistance is more than the smallest value for a single molecule but very close to it. This shows the combined effect of the individual interfacial thermal resistances. The presence of different

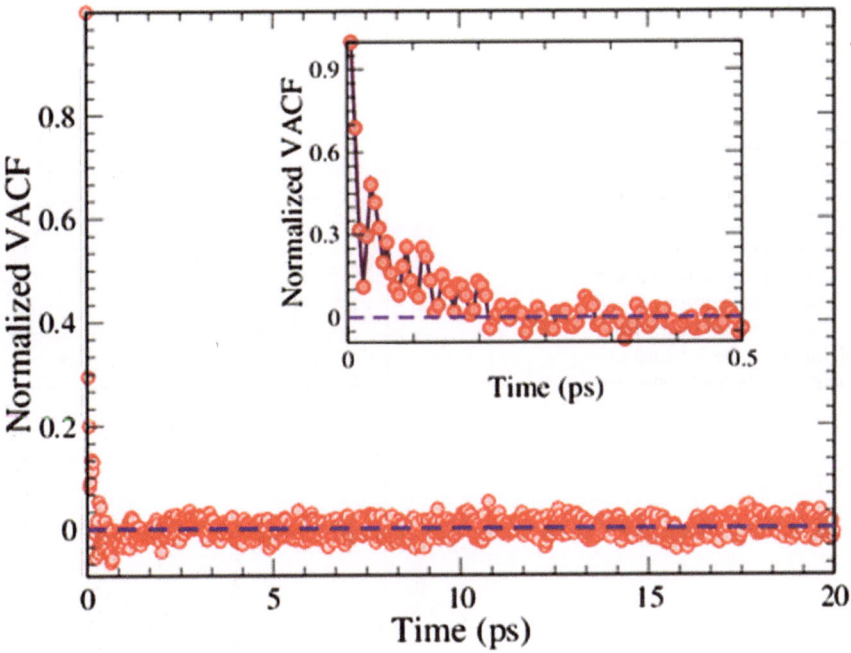

Fig. 3.6 Velocity correlation function for carbon atoms in n-heptane molecules

types of chains, with different sizes and orientations, enables the individual molecules to pack more efficiently and wrap more effectively around the nanotube.

But due to the presence of all types of chains, the effective behavior of some chains to wrap around the nanotube is countered by the ineffective behavior of other chains. So as a system, mixtures are not as efficient as the neat solvent (in which the hydrocarbon chains with the smallest size is the sole species) for minimizing the interfacial resistance. However, the mixture properties allow for a value of the Kapitza resistance that is similar in magnitude to the neat solvent (in which the hydrocarbon chains with the smallest size are the sole species).

Energy Transfer Mechanism

Space time correlations can be used to study the dynamic properties of a system. To study the energy transfer mechanism in a system, vibration spectrum of the atoms is analyzed. Vibration spectrum allows a physical insight into frequency modes of the systems. The vibration spectrum is generated by taking the Fourier spectrum of the velocity autocorrelation function of atoms under consideration. Figure 3.6 shows the temporal variation of normalized velocity autocorrelation function (NVACF) of carbon atoms of n-heptane molecules. The normalized

Fig. 3.7 Vibration spectrum for *n*-heptane (*top*), *n*-tridecane (*middle*), and *n*-nonadecane (*bottom*). The vibration frequencies have been marked for van der waals interaction (*close up*, in *insert*), and for C–C and H–H atoms. Notice the decrease in vibration amplitude at about 1–2 THz for *n*-tridecane and *n*-nonadecane corresponding to van der waals interaction highlighting the increase in interfacial thermal resistance

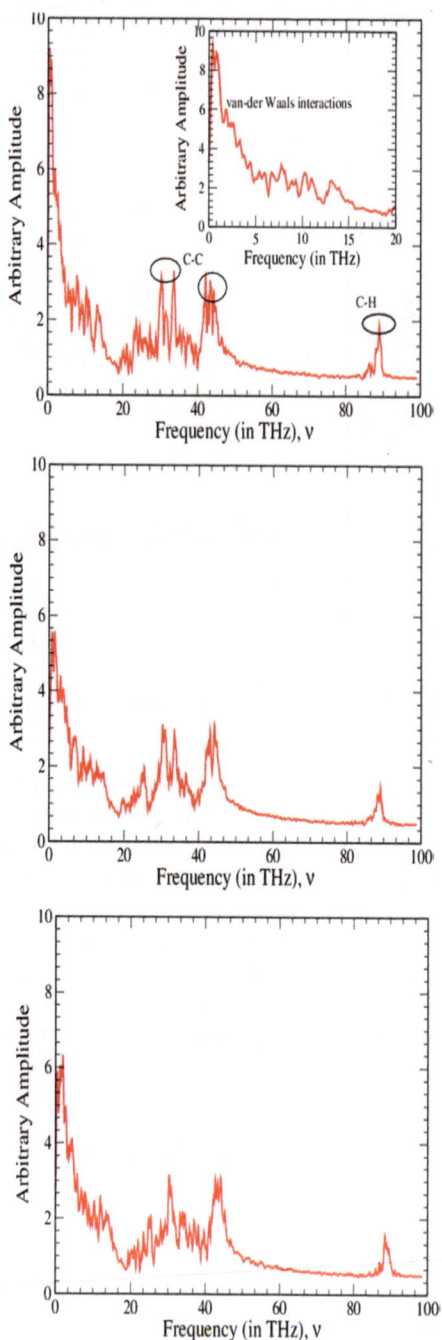

velocity correlation function starts with a value of one and fluctuates around a zero value. As the simulation is run for a very long time, the NVACF becomes zero indicating that equilibrium has reached. The atomic vibration spectrum of *n*-heptane, *n*-tridecane, and *n*-nonadecane is shown in Fig. 3.7. The peak at around 90 THz corresponds to the C–H vibration modes (Bykov 1958; Lifson and Stern 1982). The peaks around 45 THz and less correspond to the C–C stretching and other modes of vibration such as angle change, dihedrals, stretch–stretch. (Derreumaux et al. 1993; Lii and Allinger 1989). The peaks corresponding to small frequencies varying from 0.2 to 5 THz, enhanced in insert, correspond to the non-bonded van der Waals interactions. These frequencies correspond to non-bonded interaction within liquid atoms as well as between carbon atoms of liquid and carbon nanotube.

The spectrums peak frequencies are identical for all the three cases as expected owing to the fact that the vibrating atoms in all these hydrocarbons are C–C and C–H. As stated above, the low frequencies correspond to weak van der Waals coupling between the atoms. The energy transfer occurs when the atoms vibrate in resonance with each other. In the hydrocarbons, the van der Waals vibration frequency range is 0.4 THz to approx 5 THz. These frequencies correspond to the squeezing, in-plane twisting due to bond bending, and in-plane longitudinal due to bond stretching (Dresselhaus and Eklund 2000); hence, they are responsible for energy transfer from nanotube to surrounding molecules. Important information can be derived from the amplitude of the vibrations in the low-frequency range. Around 1 THz, the amplitude of vibrations is for *n*-heptane followed by *n*-nonadecane and *n*-tridecane. This decrease corresponds to lower rate of energy transfer and hence more interfacial resistance at the interface. The calculation of interfacial thermal resistance in Fig. 3.5 validates this conclusion.

Summary

From the results, it is clear that the interfacial thermal resistance strongly depends upon the chemical composition of the fluid (solvent phase). The dependence arises from the variations in the vibration frequencies of the atoms in the fluid molecules. Interfacial thermal resistance also depends on the molecular structure and varies with the length of the polymer chains, their isomers, and mixtures. The variation depends on the effectiveness of different chemical structures for their ability to wrap around the nanofin. So fluids with small and branched chains have low interfacial thermal resistance compared to long chains. In mixtures, the presence of structures of different sizes and orientation enables the individual molecules to wrap around the nanotube more effectively and thus have resistance closer to chain with the smallest resistance. Regarding the energy transfer mechanism, it is deduced that the energy from the carbon nanotubes to the fluid is transferred due to the presence of weak van der Waals interactions. The coupling of the low-frequency modes in both the nanotube and the fluid (i.e., the molecular vibrations in resonance) is the primary vehicle for transfer of thermal energy. These low-frequency vibration modes correspond to the non-bonded interactions in fluid molecules as well as squeezing and in-plane twisting due to bond bending, and in-plane longitudinal modes due to bond stretching in the nanofin.

References

Bykov GV (1958) Valence-vibration frequencies and electronic bond charges for CH bonds in hydrocarbons. Russ Chem Bull 7:743–745

Derreumaux P, Dauchez M, Vergoten G (1993) The structures and vibrational frequencies of a series of alkanes using the SPASIBA force field. J Mol Struct 295:203–221

Dresselhaus MS, Eklund PC (2000) Phonons in carbon nanotubes. Adv Phys 49:705–814

Huxtable ST, Cahill DG, Shenogin S, Xue L, Ozisik R, Barone P, Usrey M, Strano MS, Siddons G, Shim M et al (2003) Interfacial heat flow in carbon nanotube suspensions. Nat Mater 2:731–734

Lifson S, Stern PS (1982) Born-oppenheimer energy surfaces of similar molecules: interrelations between bond lengths, bond angles, and frequencies of normal vibrations in alkanes. J Chem Phys 77:4542–4550

Lii JH, Allinger NL (1989) Molecular mechanics. The mm3 force field for hydrocarbons. 2. Vibrational frequencies and thermodynamics. J Am Chem Soc 111:8566–8575

Yu CJ, Richter AG, Datta A, Durbin MK, Dutta P (2000) Molecular layering in a liquid on a solid substrate: an x-ray reflectivity study. Phys B 283:27–31

Chapter 4
Nanofins: Implications

Abstract The nuances and implications of the nanofin effect are summarized in this chapter. Future directions are identified to address the open questions in the literature pertaining to the nanofin effect.

With the latest development in the high-performance, miniaturized electronic devices comes the need to remove large amounts of heat from small areas. Pool boiling is a very efficient mode for heat transfer. Experiments have been performed to enhance the heat transfer in pool boiling by modifying the boiling surface. Recently, carbon nanotubes have attracted a great deal of attention owing to their excellent thermal, mechanical, and electrical properties. Experiments performed using carbon nanotubes as nanofins on the boiling surfaces have shown enhancement in the heat transfer. But, the enhancement level is not as high as expected from the nanotubes, especially because silicon nanofins are found to result in higher pool boiling heat flux values. This degradation in the performance of the carbon nanotubes is attributed to the presence of a larger interfacial thermal resistance between the carbon nanotubes and the surrounding fluids.

Similarly forced convection experiments involving the flow of nanofluids in microchannels were used to conclusively demonstrate the formation of nanofins. It was shown that the interfacial effects dominate the heat transfer for nanofluids, while the bulk fluid properties play a secondary role. Also, the nanofin effect was shown to affect nanofluid properties such as specific heat capacity (and possibly thermal conductivity as well as rheological behavior).

Results from noncontinuum models—such as nonequilibrium molecular dynamic simulations—were presented in this book as an example for the procedure to estimate the interfacial thermal resistance between a nanofin and solvent phase. The Kapitza resistance is the dominant parameter that affects the effectiveness of a nanofin. The simulation strategy is based on the lumped capacitance method where the temperature difference between the nanofin and the surrounding fluid decays exponentially. The simulation domain consists of a nanofin placed at the center of the simulation box surrounded by fluid molecules. The system is

N. Singh and D. Banerjee, *Nanofins*,   65
SpringerBriefs in Thermal Engineering and Applied Science,
DOI: 10.1007/978-1-4614-8532-2_4, © The Author(s) 2014

brought to a base temperature, and then the temperature of the nanofin is instantaneously raised to a higher value. As the system relaxes under constraints of constant energy condition, the energy is transferred from the nanofin to the fluid molecules. From the temporal variation of the temperature difference between nanofin and matrix molecules, the value of interfacial thermal resistance was obtained. Different types of fluid molecules (and their mixtures) were considered to elucidate the effect of various chemical and structural properties on the resultant values of the interfacial thermal resistance. It may be noted that the simulation results that were discussed in this book were restricted to single-walled CNT (SWCNT), while the experiments were performed using multi-walled CNT (MWCNT).

From the results, the following can be concluded:

- The liquid molecules near a crystalline surface are more ordered with a higher density than the bulk liquid and exhibit properties that are akin to that of a solid. The density of the liquid near the carbon nanotube peaks to a high value followed by multiple oscillations and subsequently reaches the bulk liquid density value at a few atomic distances away. This trend is observed in all the simulations shown in this book. Hence, it is expected that along with density values, other property values (such as thermal conductivity, specific heat capacity, chemical concentration of different species) and the rheological properties will be different in the compressed phase formed by the adsorbed layer of molecules on the surface of the nanofin or nanoparticle.
- The chemical concentration gradients induced locally by the presence of the nanoparticle or nanofin in turn can lead to concentration-mediated heat transfer (such as thermophoresis). Also, the chemical concentration of the different species in the compressed phase may be different (or vary spatially and temporally) than that of the bulk of the solvent phase.
- In addition, the compressed phase may have a semi-solid structure and can display non-Newtonian behavior, while the bulk of the solvent phase may be Newtonian (i.e., before the addition of the nanoparticles or before coming into contact with the surface nanofins). The semi-solid/semi-liquid compressed phase can be approximated by an atomic structure that is akin to the solid phase of the solvent that is close to the melting point. Hence, the semi-solid phase may exist in a state close to that observed in the solid phase of the solvent at the melting point—even though the operating temperature is far higher than the melting point.
- The semi-solid structure of the compressed phase can mimic the crystal lattice structure of the nanoparticle (which can be different from the lattice structure of the natural solid state—i.e., the frozen solvent). The material properties of the compressed phase can therefore vary depending on the material structure of the nanofin. The variation of the material property values of the compressed phase (and dependent parameters such as Kapitza resistance) as a function of the lattice structure that it mimics and the chemical composition of the different

species (in the compressed phase and the bulk phase of the solvent) is currently unknown in the materials science literature.

- The interfacial resistance is found to be dependent on the chemical composition of the fluid surrounding the nanotube. To study the effect of chemical composition, interfacial resistances between the nanotube and common coolants, water, ethyl alcohol, and 1-hexene were studied. From these simulations, it can be observed that the atomic properties affect the interfacial resistance due to the difference in the interaction potential between the atoms of the fluid and the nanofin. The potential depends upon the oxidation state of the atoms. So, an oxygen atom in water will have different interaction attributes than an oxygen atom in ethyl alcohol (i.e., while interacting with carbon atoms of the carbon nanotube). This change in the interaction potential affects the resultant value of the interfacial resistance between the fluid molecules and the nanofin.
- The interaction between the nanofin and surrounding fluid also depends upon the ability of the fluid to wrap around the nanofin. Coolants used in avionics systems consist of olefins (hydrocarbons) which typically exist as mixtures of monomers, dimers, trimmers, etc. Molecules with short chains can effectively wrap around the nanotube, but those with long chains cannot due to stearic hindrance. N-heptane and its dimer and trimer (n-tridecane and n-nonadecane, respectively) are considered in this study for investigating the effect of polymer chains on interfacial thermal resistance. It was shown that the small chains have lower interfacial resistance than longer chains owing to their ability to effectively wrap around the nanotube. Also, the equilibrium structure of the molecule chain affects the Kapitza resistance.
- The isomers of a hydrocarbon chain can have different molecular structures (i.e., straight chain and branched chain). Since the structure changes the effectiveness of the molecule to wrap around the nanotube, interfacial resistance for branched-chain isomers will be different when compared to that of the straight-chain isomers. Also, the manufacturing processes for these hydrocarbons produce a mixture of polymer and isomers (i.e. dimers, trimers, etc.). To study the effect of isomers and their mixtures on interfacial resistance, polymer chains and isomers of n-heptane are considered. Simulations indicate that each of the isomers can yield different values for interfacial resistance—for the same nanofin. The resistance values depend upon the effectiveness of the molecules due to their chemical structure—to wrap around the nanotube. Also, the mixtures possess lower interfacial resistance values due to more effective packing (lower stearic hindrance) of the solvent molecules around the nanofin.
- To ascertain the transport mechanism for heat transfer from the carbon nanotubes to the surrounding molecules, vibration spectrum analysis of carbon atoms of n-heptane, n-tridecane, and n-nonadecane and isomers and mixtures of n-heptane were performed. From the vibration analysis, it is observed that the weak van der Waal's interactions are responsible for the transfer of thermal (molecular vibration) energy from the nanofin to the solvent molecules. This is also apparent from the snapshot of the system at equilibrium (Fig. 3.2), as there is a gap of about 3 Å between the carbon nanotube and the fluid molecules due

to weak van der Waal's interactions. Thermal energy can be transferred only through these low-frequency vibrating modes from the heated nanofins to the fluid molecules. The fluid molecules resonate at a lower frequency—hence, they can only couple with the nanofins only for the low-frequency modes, while there is a huge impedance mismatch for the high-frequency modes that are traveling through the nanofin. The radial breathing modes (RBM) of CNT are at a much higher frequency and therefore are less likely to be effective in transferring thermal energy to the molecules in the fluid. From the vibration spectrum analysis of carbon nanotube, it is observed that squeezing (in-plane) twisting modes due to bond bending and in-plane longitudinal modes due to bond stretching are responsible for the low-frequency vibrating modes. So, these modes are responsible for transferring energy from the carbon nanotube to the fluid molecules.

Several other implications and nuances can be derived from these discussions. For example, it is proposed (Evans et al. 2008 and Prasher et al. 2005) that the thermal conductivity of a nanofluid is maximized when the diameter (D_{np}) of the (spherical or cylindrical) nanoparticle that is added into the fluid is restricted to a certain critical size depending on the thermal conductivity of the neat solvent (k_f). The critical size ($D_{np,c}$) can be expressed as a function of the Kapitza resistance (R_k) between the nanoparticle and the fluid as follows:

$$D_{np,c} = 2R_k k_f \tag{57}$$

On the other hand, the specific heat capacity is maximized as the size of the nanoparticle is reduced—preferably to less than 6 nm (and the minimum size required is 10–20 nm for the nanofin effect to be perceptible). For an organic nanofin, such as CNT in water (or organic refrigerant), the diameter of the nanoparticle for maximizing thermal conductivity is estimated to be 70–110 nm (which clearly indicates that an MWCNT will perform better than SWCNT in maximizing heat transfer to a fluid). On the other hand, a SWCNT (or MWCNT) or graphene or nanodiamond with diameter less than 6 nm will perform better in maximizing the specific heat capacity of the nanofluid. Jo (2012) reported that functionalizing the nanofins with carboxyl groups and amine groups (i.e., groups that can aid hydrogen bonding) can help to reduce the critical diameter of the nanoparticles. For example, for CNT, the critical diameter was reduced from 110 to 50 and 60 nm when functionalized with carboxyl group and amine group, respectively. Reducing the critical diameter can aid in the development of "designer nanofluids" where multiple properties (such as thermal conductivity and specific heat capacity) can be enhanced simultaneously for various applications such as for heat transfer fluids (HTF) and thermal energy storage (TES). However, chemical functionalization of nanofins can come with its own basket of drawbacks—such as compromised reliability (chemical stability), reduction of operating temperature range, variation in properties during batch synthesis, etc.

To summarize, the behavior of a nanoparticle that is in contact with a fluid is termed as the "nanofin effect". This is a complex phenomenon that not only affects the total thermal resistance (and the Kapitza resistance) but is also observed to affect thermal capacitance (thermal energy storage in the system) and thermal impedance (time constant of the system) due to fluctuations in the local value of density and chemical concentration of the different species in the solvent phase (i.e., the fluid molecules) that are in the vicinity of the nanofin/nanoparticle. The nanofin effect is found to depend not only on the thermophysical properties but also on chemical composition and chemical structure of both the nanofin and the solvent (fluid), the extent of the compressed phase that is formed on contact with the nanofin, which also includes the effects of local chemical concentration of the different species in the compressed phase and the chemical concentration gradient induced locally by the presence of the nanofin/nanoparticle (compared to a different composition in the bulk of the solvent phase or the fluid phase that is far away from the nanoparticle surface). The fluid material properties are found to be a more dominant factor than the material property of the nanoparticle itself, in determining the behavior of a nanofin (be it in a nanocoating or in a nanofluid). More specifically, the property of the fluid in the frozen state that is at a temperature close to the melting point (i.e., the solid-state property of the solvent near the melting point of the solvent material) is the more dominant parameter that affects the thermal conductivity and specific heat capacity of the nanofluid. Effects such as formation of percolation networks by the nanoparticle as well as by the compressed phase have not been discussed in this book—since they are still a developing branch in the study of nanofins, as well as nanofluids and nanocoatings.

Future Direction

This work focused on the use of a (5,5) armchair carbon nanotube as a nanofin. The effect of nanotubes of other chiralities, chiral and zigzag, and diameters on the interfacial thermal resistance was not considered. It has been established that the vibration spectrum of the nanotubes does not strongly depend upon the chirality of the carbon nanotube. So, the basic modes of energy transfer from carbon nanotube to the fluids remains the same. But due to the presence of large diameter, the effectiveness of polymer chains to wrap around the nanotubes will change. The polymer chains will be more effective in wrapping around the nanotube of larger diameters, so it is expected that the interfacial resistance may be less for large-diameter nanotubes. Further studies are needed to enumerate the effect of large diameter of carbon nanotubes on the interfacial thermal resistance for hydrocarbon chains, their isomer, and mixtures.

Also, in this work, it is assumed that the nanotube is free to vibrate without any bonding restrictions. But, this will not be true for the certain length of the nanotube near the end bounded on the silicon substrate. Even though this length is of the order of few angstrom and will not affect the bulk interfacial resistance of the nanotube, a true nanofin analysis of the nanotube can be done with one end of the nanotube fixed on a silicon substrate. To do this analysis, a nanotube of a length of few hundred nanometers to micron has to be considered. This size of the system is prohibitive for doing the molecular dynamics simulation with the current computing power available but may be feasible in the near future with rapid progress in computational capabilities of commercially available systems.

Studies have shown that the presence of a large number of defects changes the properties of the nanotubes. The nanotubes deposited on the silicon substrates using the chemical vapor deposition (CVD) process are not defect free. So, their ideal thermal transport mechanics will be disturbed by the presence of these defects. This will also affect the interfacial thermal resistance as the vibration modes of the nanotube will change. Therefore, the effect of defects on the interfacial thermal resistance needs to be studied to explore the more realistic simulations corresponding to the results obtained from the experimental data reported in the literature.

In addition, the simulation results that were discussed in this book were restricted to single-walled CNT (SWCNT), while the experiments were performed using multi-walled CNT (MWCNT). Simulation tools that are amenable for tackling MWCNT with large length (greater than 5 microns) need to be developed. However, the trends and conclusions presented in this book that are based on simulations using SWCNT are also expected to hold for long and thick nanofins.

References

Evans W, Prasher R, Fish J, Meakin P, Phelan P, Keblinski P (2008) Effect of aggregation and interfacial thermal resistance on thermal conductivity of nanocomposites and colloidal nanofluids. Int J Heat Mass Transfer 51:1431–1438

Jo B (2012) Numerical and experimental investigation of organic nanomaterials for thermal energy storage and for concentrating solar power applications. Dissertation, Texas A&M University, College Station

Prasher R, Bhattacharya P, Phelan PE (2005) Thermal conductivity of nanoscale colloidal solutions (nanofluids). Appl Phys Lett 94:025901